GEOLOGY AND MINERAL RESOURCES OF WEST AFRICA

GEOLOGY AND MINERAL RESOURCES OF WEST AFRICA

J.B.WRIGHT

editor and principal author
Department of Earth Sciences, The Open University

with contributions from

D.A.Hastings   W.B.Jones

H.R.Williams

London
GEORGE ALLEN & UNWIN
Boston   Sydney

Softcover reprint of the hardcover 1st edition 1985

George Allen & Unwin (Publishers) Ltd,
40 Museum Street, London WC1A 1LU, UK

George Allen & Unwin (Publishers) Ltd,
Park Lane, Hemel Hempstead, Herts HP2 4TE, UK

Allen & Unwin Inc.,
8 Winchester Place, Winchester, Mass. 01890, USA

George Allen & Unwin Australia Pty Ltd,
8 Napier Street, North Sydney, NSW 2060, Australia

First published in 1985

British Library Cataloguing in Publication Data

Geology and mineral resources of West Africa
1. Geology – Africa, West
I. Wright, J. B.
556.6   QE339.W4

ISBN 978-94-015-3934-0   ISBN 978-94-015-3932-6 (eBook)
DOI 10.1007/978-94-015-3932-6

Library of Congress Cataloging in Publication Data

Wright, J. B.
  Geology and mineral resources of West Africa.
Includes bibliographies and index.
1. Geology – Africa, West.   2. Mines and mineral
resources – Africa, West.   I. Title.
QE339.W46W75 1985   556.6   84-28253

Set in 10 on 11 point Plantin by Paston Press, Norwich
and

Preface

In this text, attention is focused mainly on those countries in western Africa lying south of the Sahara, that is, between about 5°N and 15°N, and westward of about 15°E. Parts of the region as far north as about 20°N are considered from time to time, for purposes of correlation and continuity. The map on p. xiii indicates the approximate extent of the coverage.

The principal aim is to provide a broad view of West African geology as a whole, for undergraduates who are studying for honours degrees in geology and who already have an understanding of basic geological principles. It is increasingly important that geologists working in this region should see it as made up of geological 'provinces' which transcend national boundaries.

As economic geology is of central importance to many developing countries, the relationship of economic mineral deposits to the different geological 'provinces' is stressed, and some attempt has been made to account for the occurrences.

The bibliography at the end of each chapter contains references to published source materials and selected items can be used as a further reading list. The bibliographies will, in general, not contain references to the publications of national geological surveys. These are normally too detailed for the general reader, but they are readily available to those seeking further information in the countries concerned. As different authorities place geological boundaries in different places, maps from different sources do not always agree in detail.

It was decided not to cite references specifically in the text, for two main reasons. First, it often interrupts the narrative flow, particularly for the student reader. Secondly, students in West Africa often find it very difficult to get access to the original literature, chiefly because of the straitened financial circumstances of so many universities. Where such literature is accessible, however, it is to be expected that teachers and lecturers will know of it and will be able to acquaint their students with it, where necessary.

A glossary of terms is provided at the end of the volume, and there is a summary at the beginning of each chapter.

This book is dedicated to the many colleagues and students with whom we have worked in West Africa and who have stimulated and encouraged our teaching and research in various ways. We hope also that it may help the work of international organizations such as AGID, CIFEG and UNESCO to encourage the growing trend towards geological co-operation and correlation between different countries in West Africa.

Responsibility for the text and illustrations rests chiefly with the principal author. The contributing authors have supplied written material and additional information, based on their extensive experience of teaching and research in West Africa. They have also provided much valuable editorial advice during the various stages of preparation, but they are not responsible for errors, omissions, or other inadequacies.

The authors regret that there was insufficient time to incorporate into the text the renaming of Upper Volta as Burkina Faso. There is no political implication in this omission, and we trust that no reader will be offended.

J. B. Wright
D. A. Hastings
W. B. Jones
H. R. Williams

Reader survey Please help us to improve future editions of this book by completing the brief questionnaire at the back of the book.

Acknowledgements

Figure 2.3 reprinted from B. F. Windley 1977. *The Evolving Continents*. London: Wiley. © 1977 B. F. Windley, by permission of the author and John Wiley & Sons, Ltd; H. R. Williams (2.4a); Figures 3.1, 3.2 and 3.5 reprinted by permission from *Nature* 272, 440–2. Copyright © 1978 Macmillan Journals Limited; Figures 3.4a&b reprinted by permission from *Nature* 282, 606–9. Copyright © 1979 Macmillan Journals Limited; Elsevier Science Publisher (5.1); Figure 5.2 reprinted by permission from *Nature* 292, 123–8. Copyright © 1981 Macmillan Journals Limited; R. Black (6.3); W. S. Pitcher (7.1); K. S. Burke (10.3); P. Lehner, P. A. C. de Reuter and the American Association of Petroleum Geologists (11.7); J. E. Ejedawe and the American Association of Petroleum Geologists (12.1); R. Black and M. Girod (13.1a&b); Figure 16.2 reproduced by permission from *Nature* 277, 152–3. Copyright © 1979 Macmillan Journals Limited.

Contents

List of tables

Contributors

D. A. HASTINGS

Adjunct Principal Geophysicist, Ghana Geological Survey *now at* EROS Data Center, Sioux Falls, South Dakota, USA.

W. B. JONES

Phillips Petroleum Company Europe–Africa, The Adelphi, John Adams Street, London

H. R. WILLIAMS

Department of Geological Sciences, Brock University, St. Catharines, Ontario, Canada.

General study guide

Following the introductory chapter, which reviews the geological setting of West Africa within the continent as a whole, there are four main parts to this book. The first part deals with the basement of ancient metamorphic and granitic rocks, which is undoubtedly the subject that students generally find most difficult to understand. Some essential concepts are introduced and explained first, with specific reference to West Africa. The distribution of rock types in the basement forms a remarkably consistent pattern throughout the whole region, irrespective of the age of the rocks concerned. Three main age provinces are recognised and described in turn: Archaean (*c.* 2500 million years, written 2500 Ma), Proterozoic (*c.* 2000 Ma) and Pan African (*c.* 500 Ma). The relevant chapters contain sections on the rocks themselves, their structural, metamorphic and (where discernible) stratigraphic relationships, their geochronology and their economic mineral potential.

The second part deals with the late Proterozoic to Mesozoic–Tertiary sedimentary sequences that occupy large shallow inland basins on the metamorphic and granitic basement and deeper coastal basins along the continental margin. Individual chapters in this part describe the basins in turn, beginning with the oldest. Each chapter outlines the distribution of rock types in the sequence, their facies variations in relation to the palaeogeography of the basin, and their economic mineral potential.

The third part deals with the anorogenic igneous activity of mainly post-Precambrian (Phanerozoic) age that occurs throughout West Africa. Some of this activity is demonstrably related to the Mesozoic separation of the African and South American continents. Individual chapters describe petrographic provinces, each characterised by rocks related in time, space and geochemistry. Both field and petrological relationships are discussed, as well as the geochronology and economic mineral potential of the rocks.

The fourth and final part examines the Quaternary deposits in relation to geologically recent tectonic events and changes in climate and geomorphology, and reviews constructional materials and engineering aspects as well as water resources.

As noted in the preface, minimal emphasis is placed on national boundaries, so that geological and structural units or provinces can be described and discussed in their entirety.

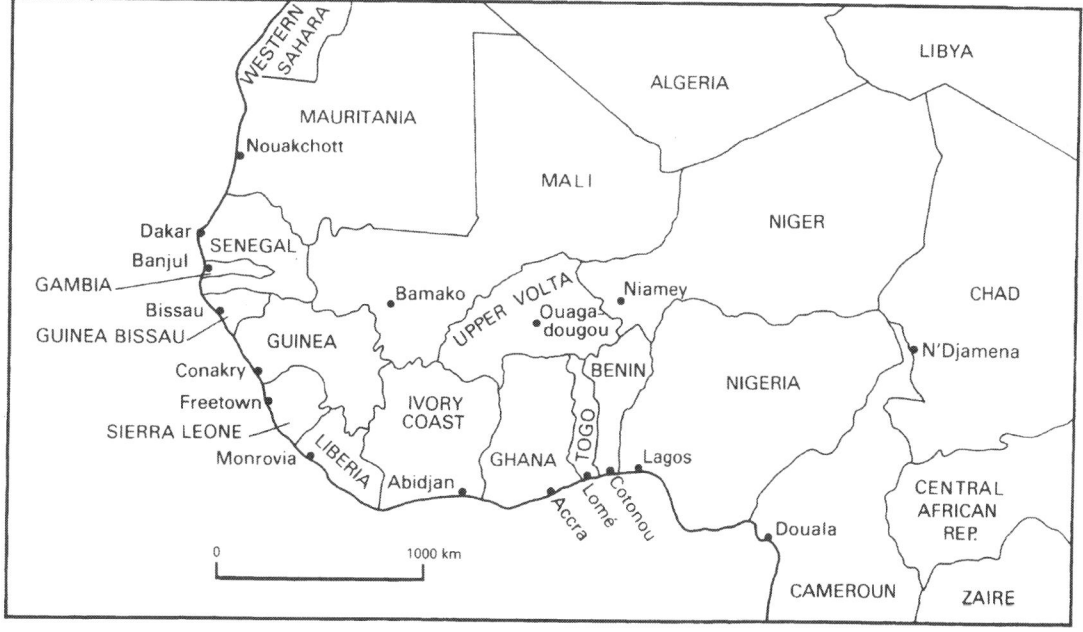

1   *The geological setting*

SUMMARY

This chapter first outlines the main geological and physiographical elements of Africa. A brief review of the major subdivisions of geological time provides a framework within which the boundaries of cratonic and non-cratonic areas of continental crust can be defined and different age provinces identified. Plate tectonic processes have probably been important in crustal evolution since the oldest rocks were formed.

The identification of age provinces can be of economic importance, because the relative abundance of mineral deposits that characteristically accompany specific associations of rocks depends to a great extent on the ages of those rocks.

The chapter then summarises the principal geological units of West Africa, with the help of a synoptic map that can be used for general reference purposes.

Geological correlation can present problems across national boundaries and these problems are increased in West Africa, where exposure is often poor and access difficult and where absolute age determinations have only recently become numerous.

1.1   The African continent

Figures 1.1 and 1.2 present the fundamental pattern of African geology, emphasising the two major features. Virtually the whole continent is composed of highly deformed metamorphic rocks and granitic intrusions. These form the surface over about half of the total area. The other half is underlain by a rather thin layer of nearly flat-lying sedimentary strata, which occupy broad shallow basins and are sharply unconformable on the often steeply inclined metamorphic rocks, often referred to by the general term **basement**.

The metamorphic and granitic rocks are almost everywhere older than about 500 million years (500 Ma), that is, they are mostly Precambrian. Most of the sedimentary rocks are younger than about 600 Ma, i.e. they are mainly of Palaeozoic and Mesozoic to Tertiary age.

There is some overlap of ages between these two main groups of rocks. In northwestern Africa, for example, some of the sediments of the Taoudeni (sometimes spelt Taoudenni) and Volta Basins are of late Precambrian age; and some of the metamorphic rocks in Mauritania and Senegal are Palaeozoic, products of the **Hercynian** orogeny. The sediments in the basins are occasionally quite strongly folded. The Benue Trough of Nigeria is perhaps the best-known example.

The only true 'mountain range' in Africa is in the extreme north-west, the Atlas Mountains of Morocco and Algeria. But these should be regarded as part of the major fold mountain chain that includes the European Alps and the Himalayas of Asia. Almost the whole of Africa has been unaffected by major mountain-building (orogenic) earth movements since the end of the Precambrian.

The metamorphic rocks and intrusive granites of the kind that form most of Africa are the products of intense deformation, regional metamorphism and partial melting that occur in the root zones of orogenic (mountain-building) belts. Such belts are zones of crustal thickening, and their lower parts are depressed as the crust attains isostatic equilibrium. The rate of erosion is proportional to height above sea level, so the upper parts of a mountain belt are relatively rapidly eroded. The crust rises to maintain isostatic equilibrium as erosion continues, though at a progressively decreasing rate, eventually exposing the high-grade rocks that formed at deep levels. There has been plenty of time for erosion to expose the root zones of the ancient orogenic belts of Africa, because the events responsible for them took place mostly in the Precambrian, more than 500 Ma ago.

Africa is predominantly a continent of plains and plateaux and intervening escarpments, the result of erosion and planation lasting many hundreds of millions of years. Topographic relief is low over vast areas. Where slopes are steep they are commonly in regions where large volcanoes have grown up or where doming and major rifting has occurred – as in much of eastern Africa.

Geological and structural mapping of the Precambrian metamorphic and granitic terranes* has

*Here, *terrane* is used to mean 'underlying geological structure' while *terrain* is used to refer to 'the visible landscape'.

1

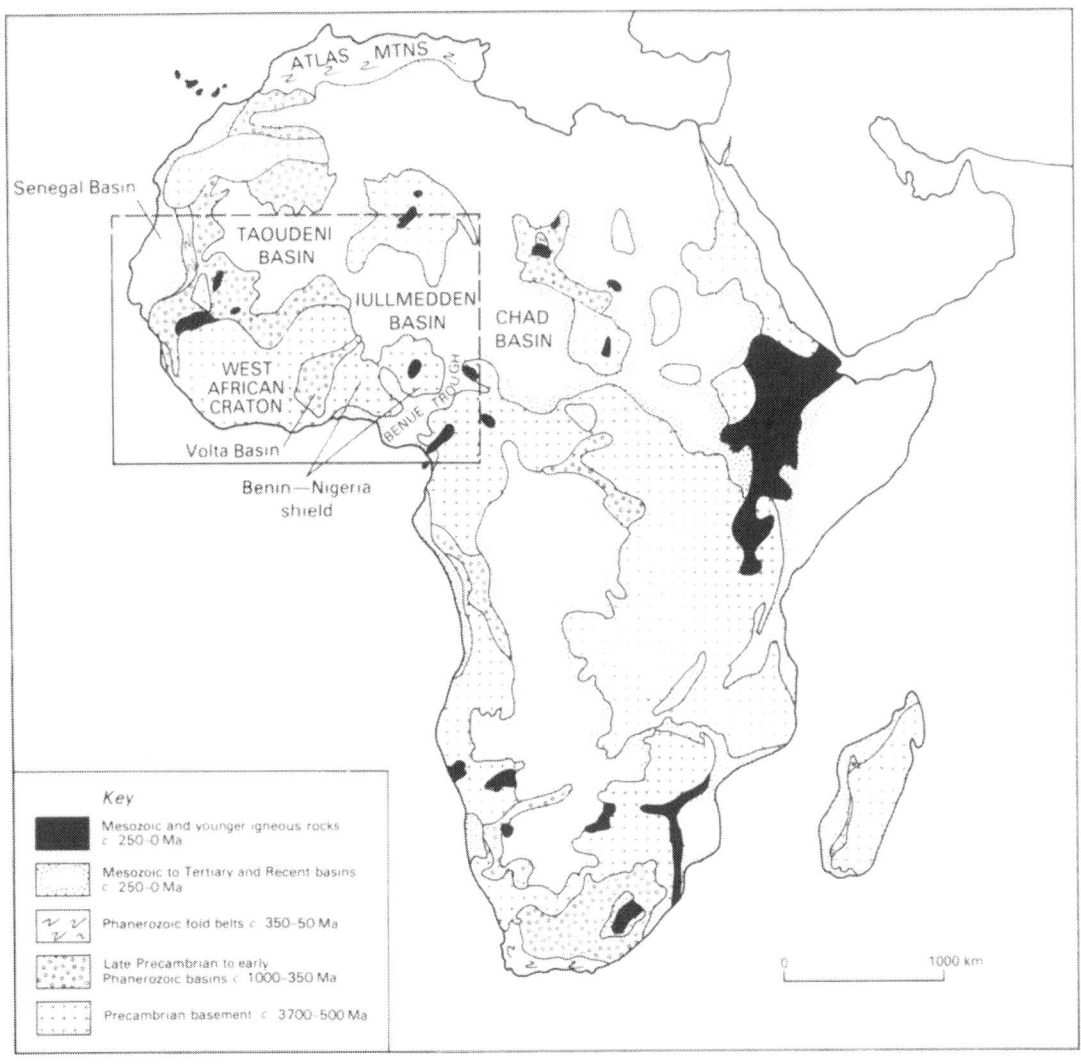

Figure 1.1 The 'basin-and-swell' structure of the African continent. The swells are underlain by older metamorphic and igneous rocks. The basins contain mostly undeformed younger sediments. The rectangle outlines the approximate area covered in this volume.

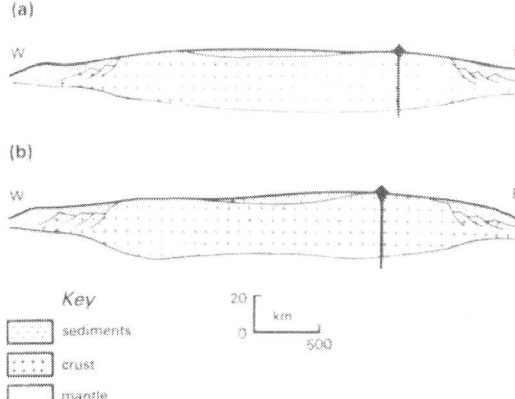

Figure 1.2 Highly schematic cross section through Figure 1.1, approximately along the Equator, with greatly exaggerated vertical scale. It illustrates two possible explanations of the 'basin-and-swell' configuration of Africa. (a) The continental crust is gently warped to form the rises (swells) and basins. (b) The swells are isostatically elevated because the crust beneath them is slightly thicker than normal. The basins are correspondingly depressed because the crust beneath them is slightly thinned. There is growing evidence that most large sedimentary basins have formed in this way. The maximum thickness of sediments in the continental basins is nowhere more than a few kilometres. The continental crust is normally 30–35 km thick. Continental crust thins by stretching and faulting at continental margins, grading into thinner and denser oceanic crust of basaltic composition.

defined the root zones of several major orogenic belts, and their ages have been determined by radioactive dating techniques. Before discussing the distribution of metamorphic and granitic rocks of different ages in Africa, it is necessary to define a basic frame of reference for geological time.

1.2 The major subdivisions of geological time

The subdivisions listed and described below have been established by worldwide correlation of radiometric age determinations and are now in general use. The following terminology is used in this text:

X million years, abbreviated to X Ma, also expressed as $X \times 10^6$ years
(the M stands for mega, meaning 10^6)
Y billion years, abbreviated to Y Ga, also expressed as $Y \times 10^9$ years
(the G stands for giga, meaning 10^9)

Archaean: All rocks older than about 2.5 billion years (2.5 Ga or 2500 Ma) belong to the Archaean. At the time of writing, the age of the oldest known rocks on the Earth is 3.8 Ga. The Earth itself is thought to be around 4.6 Ga old – so the first 800 Ma of the Earth's history are not represented in the known geological record.

Proterozoic: Rocks in the age span from about 2500 Ma to about 600 Ma belong to the Proterozoic. (The term comes from the Greek for 'early life'. Although life forms are now known to have existed in the Archaean, they become more widespread from post-Archaean time onwards.) The Proterozoic can be further subdivided into:

Upper (*c.* 1000–600 Ma),
Middle (*c.* 1800–1000 Ma),
Lower (*c.* 2500–1800 Ma).

The *Archaean–Proterozoic boundary* was originally defined as the division between rocks with signs of life preserved in them and rocks without. It is now widely believed to represent a fundamental change in the pattern of evolution of continental crust. There is evidence that the Archaean was a period of rapid growth of continental crust, during which about three-quarters of the present global extent of continental crust was formed.

The Archaean and Proterozoic are subdivisions of the *Precambrian*, which represents by far the largest proportion, about 85%, of the whole of geological time. Formerly considered to be wholly barren of fossil remains, more and more Precambrian rocks are being found with traces of life forms preserved in them.

Ages quoted for subdivisions of the Precambrian must not be regarded as rigid time boundaries. Major geological events do not necessarily happen at the same time all over the world.

Phanerozoic: The last 600 Ma or so of geological time, i.e. from Cambrian to the Present, is called the Phanerozoic. Its base (the Cambrian) is defined by the first appearance of animals with hard shells or skeletons. Fossils are therefore abundant in Phanerozoic rocks. They enabled geologists to subdivide and correlate this span of geological time worldwide, long before techniques of radiometric age measurements were developed.

The term **Infracambrian** is sometimes used to denote rocks of uncertain but probably late Precambrian age, which underlie sediments of known Cambrian age.

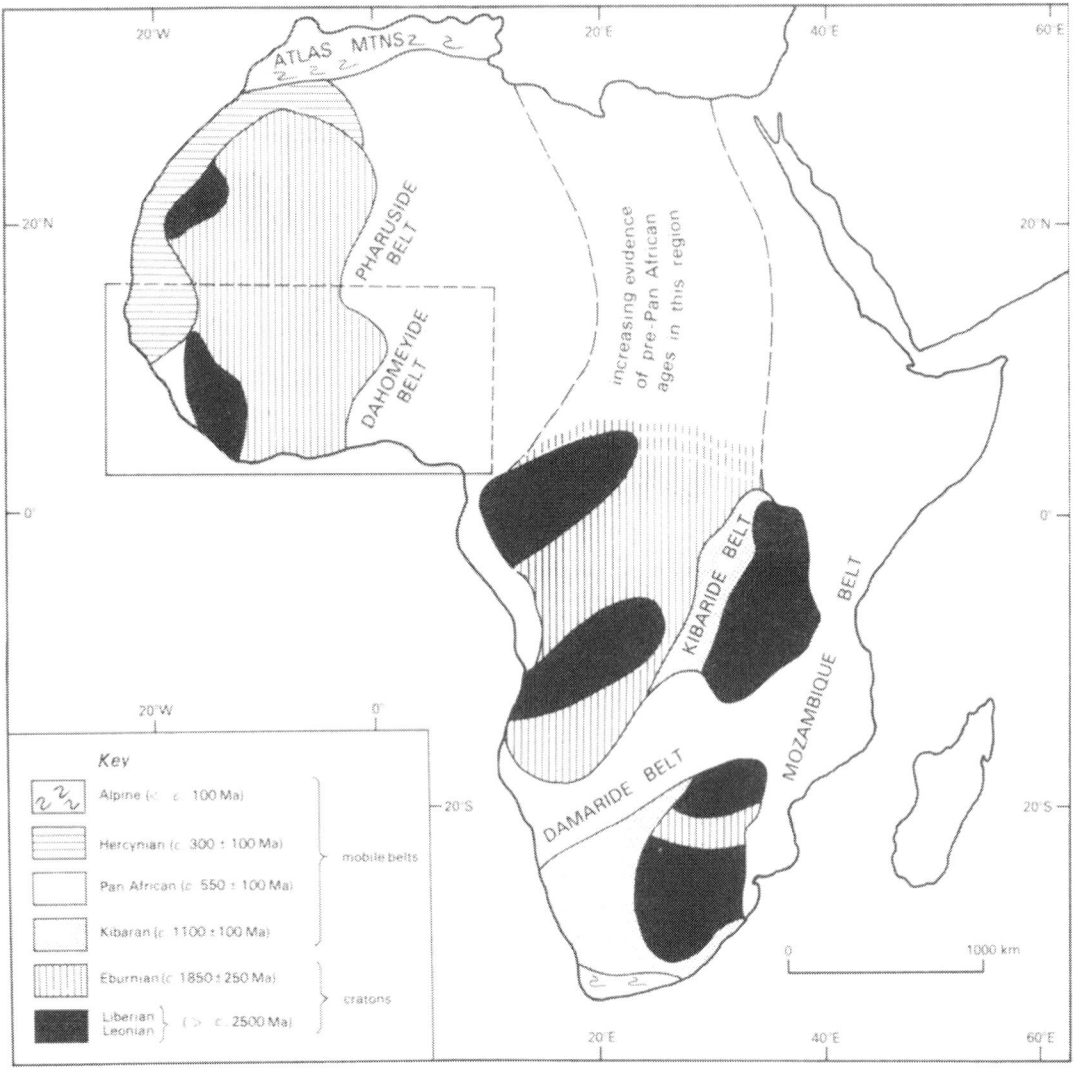

Figure 1.3 The structural framework of Africa. Sedimentary basins are not shown. The continent is made up of rocks deformed and metamorphosed during the periods of geological time listed in the key. The rectangle outlines the approximate area covered in this volume. Note that the names Liberian and Leonian are applicable to West Africa only.

In summary, then, almost the whole of Africa is made up of Precambrian rocks. The mainly Phanerozoic sedimentary basins constitute a comparatively thin cover on basement of these older rocks, which form the structural framework of the continent. It is now possible to examine that framework in more detail.

1.3 The structural framework of Africa

Figure 1.3 shows the African continent without the mainly Phanerozoic cover of nearly flat-lying sedimentary rocks in the large basins. The metamorphic and granitic rocks that make up most of the continent have been divided into two major groups:

4

(1) The **cratons** are made up of orogenic belts that are older than about 1500 Ma. They have been geologically stable – i.e. they have suffered no major orogenic deformation or metamorphism – since Lower to Middle Proterozoic times. Within the cratons are smaller **cratonic nuclei** of Archaean rocks that were deformed and regionally metamorphosed more than 2500 Ma ago and have been stable since then. The cratons are often also called **shields**.

(2) The younger **mobile belts** are made up of rocks that have undergone no orogenic deformation or regional metamorphism since the lowermost Phanerozoic, about 500 Ma ago – with the exception of those in the extreme south and north-west of the continent, which are of later Phanerozoic age. Some belts became stabilised about 1100 Ma ago, but the great majority of rocks in the mobile belt region of Figure 1.3 experienced their last major deformation and metamorphism around 500 Ma ago, in the **Pan African** event.

The ages given above are approximate, with considerable 'error limits' (Figure 1.3). This is because orogenies are commonly **diachronous** – they occur at different times in different places. It should be clear from Figure 1.3 that virtually the whole of the African continent has been tectonically stable since about 500 Ma ago. For most of Phanerozoic time it has been subjected only to **epeirogenic** movements of differential uplift and subsidence and faulting. Because of this overall stability, the whole continent is sometimes referred to as the African shield.

Each of the regions outlined in Figure 1.3 can be thought of as an **age province**, in the sense that the great majority of the rocks from that region give radiometric dates falling within a particular range of ages. The rectangle in Figure 1.3 shows that West Africa encompasses the Archaean nucleus of the West African craton, and portions of four other age provinces: a segment of the Proterozoic terrane making up the rest of the craton, and parts of the mobile Pan African and younger mobile belts that lie to the east and west of it. These age provinces are individually described in subsequent chapters.

1.4 The global context

Extensive areas of regional metamorphic rocks are known on all the other continents. The distribution of ancient orogenic belts is generally similar to that in Africa: there are older cratons with Archaean nuclei, and younger belts which have become tectonically stable at various times since the Archaean.

The extent to which seafloor spreading and plate tectonics have contributed to continental evolution has been much debated since the 1960s. What is certain is that the crust beneath the oceans is nowhere older than about 200 Ma, and the distribution of continental crust in the geological past was quite different from what it is today. For example, the age provinces of West Africa have their continuation in northeastern South America, for these two continents were once joined together.

There is less certainty about when seafloor spreading and plate tectonics began. Some authorities believe these processes became important only in the late Proterozoic, others that they were already operating in some form early in the Archaean, and still others who would place their commencement somewhere between those two extremes.

Palaeomagnetic evidence from different parts of the African continent has been interpreted as suggesting that the principal cratonic areas of Figure 1.3 were in approximately their present relative positions and orientations as long ago as 2500 Ma. If the successive regional deformation and metamorphic events that have shaped Africa resulted from the convergence and collision of lithospheric plates, this interpretation of palaeomagnetic data places constraints upon the size of ocean basins that may have opened, spread and ultimately closed between colliding continental plates: such basins could not have been more than 1000 km across at most.

The palaeomagnetic evidence is not conclusive, however, for it has also been interpreted in quite the opposite way, namely to suggest that substantial relative movements between different segments of the African continent could have occurred during Archaean and Proterozoic times. Such a view is consistent with plate tectonic models for the crustal evolution of regions as old as Lower Proterozoic in other continental areas, where less controversial palaeomagnetic data imply considerable movements of crustal segments relative to one another.

Part of the reason for conflicting interpretations of palaeomagnetic data lies in the fact that such data provide no information about movements parallel to latitude.

Another way of looking at the problem is to consider the nature of seafloor spreading and plate tectonic processes themselves. There can be little doubt that they result from convective circulation in the mantle. This is 'driven' by the Earth's internal heat, which comes from the decay of radioactive heat-producing elements. These have been decaying and decreasing in abundance through geological time, which means that the Earth's internal temperature was once higher than it is now, with steeper thermal gradients. If present-day seafloor spreading and plate tectonics are related to mantle convection, then it is at least plausible that they contributed to crustal evolution in the geological past, when rates of convection were greater than they are now. This essentially uniformitarian view is adopted here, though it may turn out not to be the correct one, because the evidence remains inconclusive and is capable of more than one interpretation.

1.5 Age provinces and mineral deposits

The recognition that continental crust is made up of belts of rocks of widely differing ages and speculations about the role of plate tectonics in the evolution of that crust are of fundamental scientific interest and importance. But they also have implications in the search for deposits of valuable minerals, and their importance will grow as the demand for raw materials increases and new deposits become more difficult to find. Metal ores and deposits of other useful materials are found in specific geological environments – petroleum in sedimentary rocks, tin in granites, and so on.

In recent years it has become increasingly apparent that the age of such environments is as important as the rock types involved. Even on the scale of the African continent it is possible to see a correlation between the occurrence of economic mineral deposits and the age of the crust. The rock *associations* appropriate to various economic mineral deposits occur in many places, but some of them tend to be more abundant in rocks of certain ages. For example, major deposits of diamonds, gold, chromite and iron ores occur in cratonic areas of Archaean to Lower Proterozoic age, whereas important copper, tin, lead and zinc ores are more commonly found in younger rocks. Age is not the only criterion, of course, and there are plenty of exceptions to these generalisations. Nonetheless, there is a strong correlation between what are known as **metallogenic** or **mineralogenic provinces** and geologically circumscribed age provinces.

The descriptions of individual age provinces are

extended to the associated mineral deposits in succeeding chapters. At this stage, two points need to be made in connection with the economic mineralisation of West Africa.

First, placer and residual deposits, such as gold and bauxite, are often the result of Tertiary–Quaternary weathering and erosion under tropical conditions. However, they are dealt with according to the age of the province in which the host rocks occur.

Secondly, the mineral deposits are often small in size and not suitable for large-scale mining enterprises. Interest in small-scale mining is increasing in many developing countries, however, and, as the industrial infrastructure in these countries grows and prospers, small low-cost mining ventures can contribute to economic development by providing key raw materials that would otherwise have to be imported.

1.6 The geological framework of West Africa

Figure 1.4 summarises the principal geological units of West Africa. It contains elements of the basin-and-swell pattern shown in Figure 1.1 as well as the distribution of Precambrian age provinces shown in Figure 1.3, delineated here in more detail.

In the west is the Guinea Rise, formed by the southern half of the West African craton, sometimes also called the Man shield. The narrow Pan African belt of the Rokelides on its western boundary is associated with some marginal reactivation of the Archaean rocks. (Sec. 2.1).

North of the craton lie Infracambrian to Lower Palaeozoic sediments of the great Taoudeni Basin, obscured in the east by continental Tertiary to Quaternary deposits. The Bové Basin in the southwest is occupied by Lower Palaeozoic sediments which interrupt the continuity of the Rokelide belt with the Mauritanide belt, which forms the eastern boundary of the much younger (Mesozoic–Tertiary) Senegal Basin.

The eastern part of the craton is overlain by Infracambrian to Lower Palaeozoic sediments of the Volta Basin, separated by the intensely thrust-faulted rocks of the Togo belt from the mainly Pan African rocks of the Togo–Benin–Nigeria swell, also known as the Benin–Nigeria shield. This is crossed by the narrow Cretaceous Bida Basin and Benue Trough, which link up with the Cretaceous to Quaternary sediments of the Niger Delta. The Pan

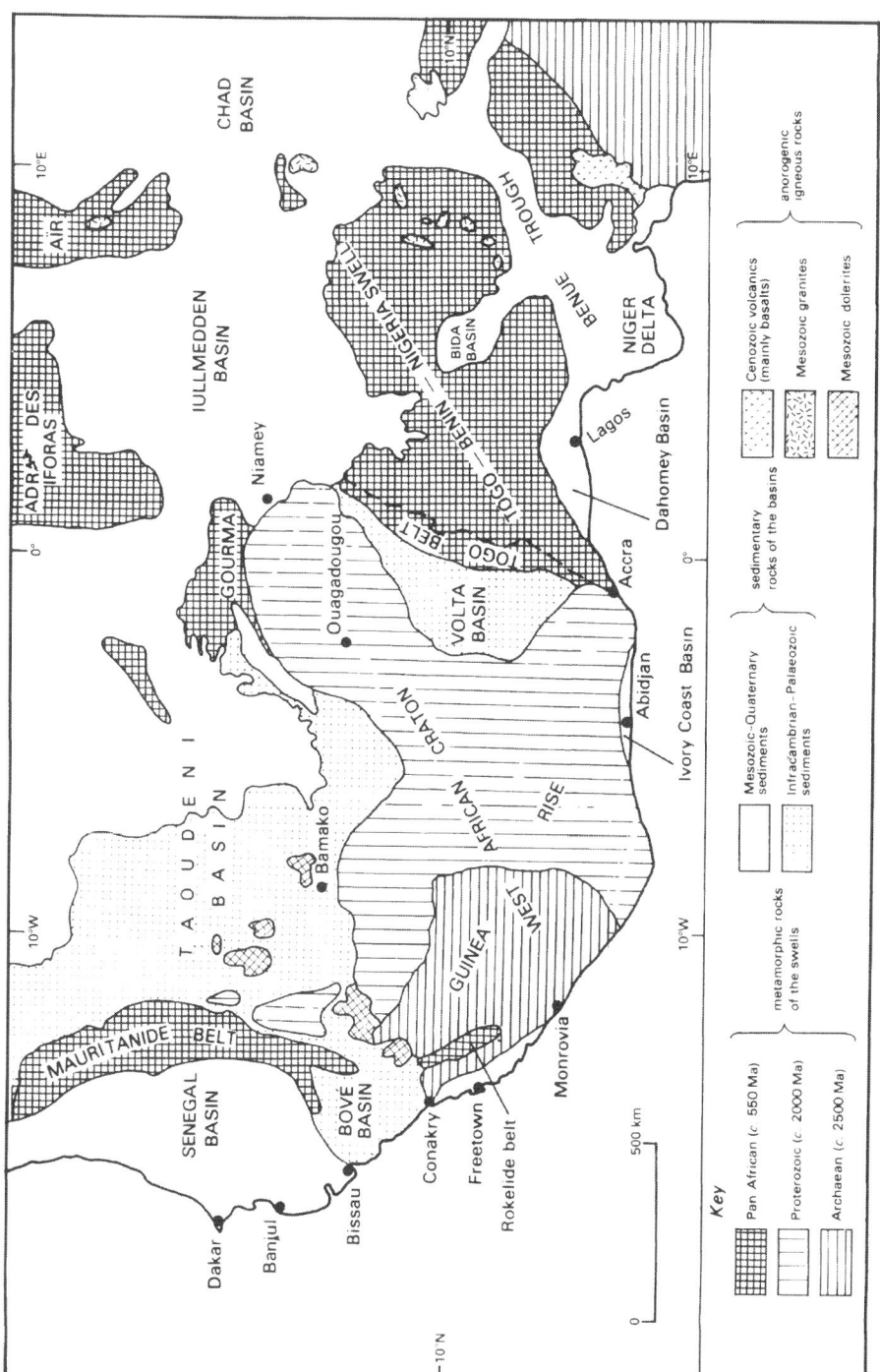

Figure 1.4 Synoptic map of the main geological units of West Africa.

Key

metamorphic rocks of the swells
- Pan African (c 550 Ma)
- Proterozoic (c 2000 Ma)
- Archaean (c 2500 Ma)

sedimentary rocks of the basins
- Mesozoic–Quaternary sediments
- Infracambrian–Palaeozoic sediments

anorogenic igneous rocks
- Cenozoic volcanics (mainly basalts)
- Mesozoic granites
- Mesozoic dolerites

CHAD BASIN

AIR

ADRAR DES IFORAS

IULLMEDDEN BASIN

GOURMA

NIGER SHELL

BIDA BASIN

BENIN-NIGERIA BELT

TOGO BELT

BENUE TROUGH

NIGER DELTA

Lagos

Dahomey Basin

Accra

Ivory Coast Basin

Abidjan

VOLTA BASIN

AFRICAN CRATON

RISE

WEST AFRICAN RISE

GUINEA RISE

Monrovia

Freetown

Rokelide belt

Conakry

Niamey

Ouagadougou

Bamako

T A O U D E N I   B A S I N

MAURITANIDE BELT

SENEGAL BASIN

BOVÉ BASIN

Bissau

Banjul

Dakar

500 km

0

10°N

10°W

0°

10°E

10°N

African rocks in the Adrar des Iforas and the Aïr, which form the northern boundary of the Mesozoic–Tertiary Iullmedden Basin, are southern extensions of the basement block of the Hoggar, and the Pan African rocks of the Gourma lie on its western boundary. Further east is the Chad Basin, mainly occupied by Quaternary sediments. Along the southern coast of West Africa are smaller narrower basins dating back to the Cretaceous, in which sediments are still accumulating at the present time.

Large dolerite sills of Mesozoic age occur in the southwestern Taoudeni Basin, and there is an important Palaeozoic to Jurassic and Tertiary province of granitic intrusions in Niger, Nigeria and Cameroun. Basaltic volcanism was widespread in the Cenozoic, mainly east of the craton.

1.7 Problems of regional treatment and correlation

One of the major problems that accompanies a regional treatment of geology is that the same rock units are given different names in adjoining countries. Another is that correlations are often different across national boundaries. Moreover, Africa has been subject to erosion for prolonged periods. Superficial deposits are widespread and weathering is deep, so that exposures are often very poor.

The problems of correlation and nomenclature of rock units are particularly acute among the metamorphic and plutonic rocks of the 'swells'. Radiometric age measurements have become numerous only since the early 1960s. Before then there was no satisfactory way of dating these rocks, because they contain no fossils, so that even their relative ages were impossible to determine with certainty, let alone their absolute ages.

As a result, age relationships published as recently as the 1960s have since proved to be incorrect. In West Africa, for instance, the rocks of the Guinea Rise are shown on older maps to be mostly younger than those of the Togo–Benin–Nigeria 'swell'. The opposite is now known to be the case.

Such radical reappraisals can lead to confusion, because names given to geological units, and originally intended to represent particular intervals of geological time, may now have meanings quite different from those they formerly had. Thus, the name **Dahomeyan** was given to the metamorphic and plutonic rocks east of the craton at a time when these were thought to be the oldest rocks of West Africa. The name **Birimian** was given to an extensive area of rocks west of the Volta Basin and originally carried the implication that these were younger than the Dahomeyan rocks belonging mainly to the Pan African mobile belt; whereas in fact, the Birimian rocks are of Proterozoic age and are part of the West African craton.

The principal reason for interpretations such as these is not hard to see: the proportion of structurally complex high-grade gneisses is much greater among the Dahomeyan rocks of the Togo–Benin–Nigeria 'swell' than among Birimian rocks west of the Volta Basin, which are mostly relatively low-grade greenstones, greywackes and phyllites. The latter were therefore assumed to be younger. At the time, this was a perfectly plausible assumption to make, but as the number of radiometric age measurements increased it soon became apparent that it was incorrect.

It is now quite clear that an Archaean gneiss or schist looks no different from its counterpart of, say, Pan African or even younger age, either in the field or in a hand specimen. Nor are there any obvious differences in metamorphic grades or in styles of structural deformation. The same applies to comparable rocks of any age anywhere in the world. As will become abundantly clear in subsequent chapters, *metamorphic grade and degree of deformation are no guide to geological age*.

Geological mapping is vitally important to establish the distribution of different rock types and their metamorphic and structural history. However, it can only establish relative ages locally (e.g. igneous intrusions) or in areas of good exposure, which are not common in West Africa, and metamorphic rocks contain no fossils. Above all, it cannot establish absolute ages. Nonetheless, radiometric age measurements are of little value unless they can be related to the results of careful and thorough geological mapping. These points should be kept in mind when reading what follows.

It is now appropriate to turn to some of the fundamental relationships that are crucial to a proper understanding of the Precambrian rocks of West Africa.

Bibliography

AGID News, no. 30, Jan. 1982. Special issue on small scale mining. Brasilia: University of Brasilia.

Black, R. 1980. Precambrian of West Africa. *Episodes* **1980**, no. 4 (Dec.).

BIBLIOGRAPHY

Bonhomme, M. G. and J. Bertrand-Sarfati 1982. Correlation of Proterozoic sediments of western and central Africa and South America, based upon radiochronological and palaeontological data. *Precambrian Res.* **18**, 171–94.

Burke, K., J. F. Dewey and W. S. F. Kidd 1976. Precambrian palaeomagnetic results compatible with contemporary operation of the Wilson cycle. *Tectonophysics* **33**, 287–99.

Clifford, T. N. 1970. The structural framework of Africa. In *African magmatism and tectonics*, T. N. Clifford and I. G. Gass (eds). London: Oliver & Boyd.

Clifford, T. N. 1972. Location of mineral deposits. In *Understanding the Earth*, 2nd edn, I. G. Gass, P. J. Smith and R. C. L. Wilson (eds). London: Artemis Press.

Furon, R. 1963. *Geology of Africa* (English edn). London: Oliver & Boyd.

Haughton, S. H. 1963. *The stratigraphic history of Africa south of the Sahara*. London: Oliver & Boyd.

Holmes, A. 1965. *Principles of physical geology*, 2nd edn. London: Nelson.

Kröner, A. 1977. The Precambrian geotectonic evolution of Africa. *Precambrian Res.* **4**, 163–213.

Lewry, J. F. 1981. Lower Proterozoic arc–microcontinent collisional tectonics in the Western Churchill Province. *Nature* **294**, 69–72.

Piper, J. D. A., J. C. Briden and K. Lomax 1973. Precambrian Africa and South America as a single continent. *Nature* **245**, 244–8.

Tarling, D. H. (ed.) 1978. *Evolution of the Earth's crust*. London: Academic Press.

Tarling, D. H. 1980. Lithosphere evolution and changing tectonic regimes. *J. Geol. Soc. Lond.* **137**, 459–67.

Part I

THE PRECAMBRIAN OF
WEST AFRICA

2 Crustal development in West Africa

SUMMARY

This chapter begins with the important concept of crustal reactivation, outlining how particular areas of continental crust can be involved in repeated orogenic events of deformation and regional metamorphism. The age of a particular geological province represents the time of the last of these events.

Regional patterns of the West African Precambrian show some striking similarities, irrespective of age. In many areas a three-fold subdivision can be recognised comprising varying proportions of (a) a high-grade granodioritic gneiss–migmatite basement, (b) elongate synclinorial belts of lower-grade supracrustal greenstones, schists and phyllites, representing original sedimentary and volcanic rocks and (c) syntectonic to late-tectonic intrusive granites. Regional structural trends are almost everywhere between N–S and NE–SW.

These fundamental patterns lead to the suggestion that the West African crust has experienced several repeated cycles of essentially similar events. The nature of those events remains an outstanding problem. There is still no consensus about such things as, for instance, the original continuity of supracrustal sediments and volcanics, the way in which they have been downfolded or downfaulted into the basement, and whether they were originally deposited on oceanic or continental crust. Repeated reactivations involving migmatisation and granitisation have progressively obliterated older supracrustal sequences, of which only remnants are now preserved, so that at least the upper continental crust retains a bulk composition near that of granodiorite. The chapter ends with a summary of the main events in the Precambrian evolution of the West African continental crust.

2.1 The concept of crustal reactivation

At first sight, the distribution of age provinces on maps such as Figure 1.3 gives the impression that the total area of continental crust has been growing steadily since the Archaean, by a process of lateral accretion. Successive orogenies appear to have added new areas of regional metamorphic and granitic rocks, thus progressively increasing the total extent of continental crust. The true picture is not quite so simple.

The ages used to define provinces such as those in Figure 1.3 give the approximate time of the last orogenic deformation and metamorphism, according to the radiometric dates yielded by the majority of the rocks in that region. In most of the age provinces, however, some rocks give ages much older than the majority, and these are often called **relict ages**. Such rocks have preserved a record of earlier regional deformation and metamorphic events. They have been through more than one orogeny.

Figure 2.1 is the same map as Figure 1.4 but without the sedimentary cover. For each age province, the list of ages includes not only the range which is given by the majority of the rocks – and therefore defines the province – but also the older ages which have been obtained from some of the rocks. For instance, rocks from the Pan African sector east of the craton have given radiometric ages recording the effects of no fewer than four previous orogenic episodes going back to the Archaean. West of the craton, on the other hand, only relatively young Pan African and Hercynian ages are recorded.

The results of geological mapping in many Precambrian regions of the world, combined with distribution of ages such as those shown in Figure 2.1, provide the basis of the statement made in Section 1.2, that most of the Earth's continental crust was formed by the end of the Archaean. Figure 2.1 provides evidence that a large proportion of the crust of what is now West Africa was already in existence 2500 Ma ago, but that much of it was affected by later orogenies. In each orogeny there was an incremental addition to the crust in the form of newly metamorphosed sedimentary and igneous rocks and granitic intrusives. But great areas of previously regionally metamorphosed older crust were also caught up in these orogenies, heated and subjected to further deformation and metamorphism on a regional scale.

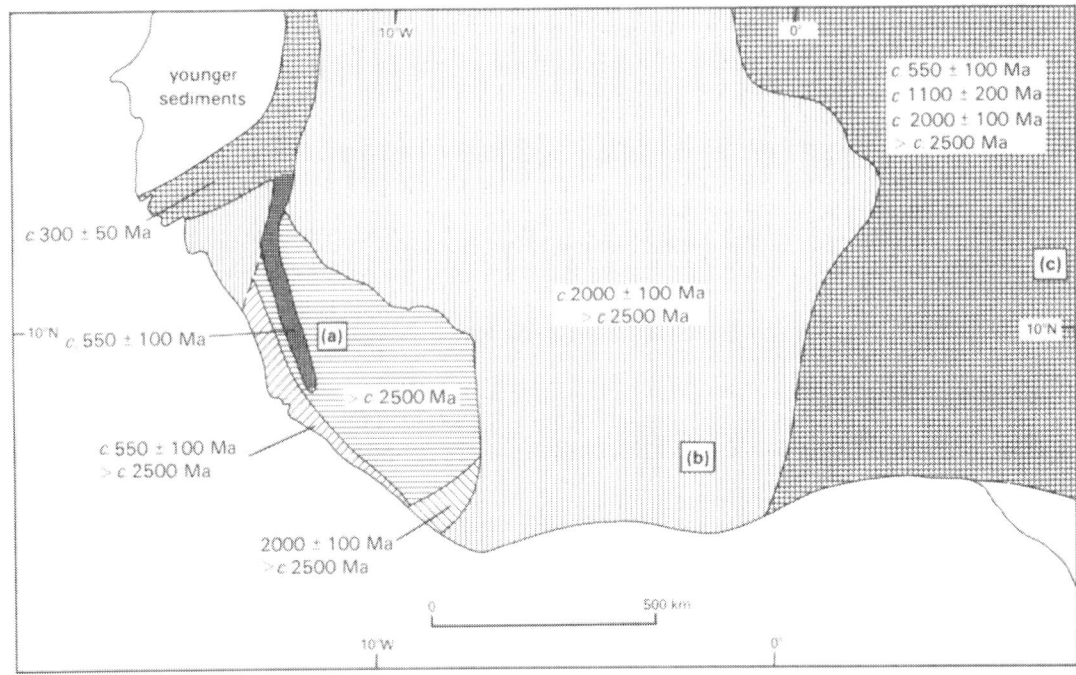

Figure 2.1 Distribution of radiometric dates in the different orogenic provinces of West Africa. The youngest age range for each province is that yielded by most of the rocks in it. Older ages are provided by a relatively small proportion of the rocks. The age ranges are approximate and will be reviewed in more detail in later chapters. Squares labelled (a) to (c) are locations for areas shown in Figure 2.4.

This process is known as **crustal reactivation** and Figure 2.2 illustrates one way in which it might be brought about: by plate tectonic processes leading to continental collision.

As the old crust is heated up and deformed once more, shear zones and faults provide pathways for water and easily mobilised constituents (notably the alkali elements, along with others such as boron) to move through the rocks. These promote isotopic, textural and mineralogical changes, often leading to

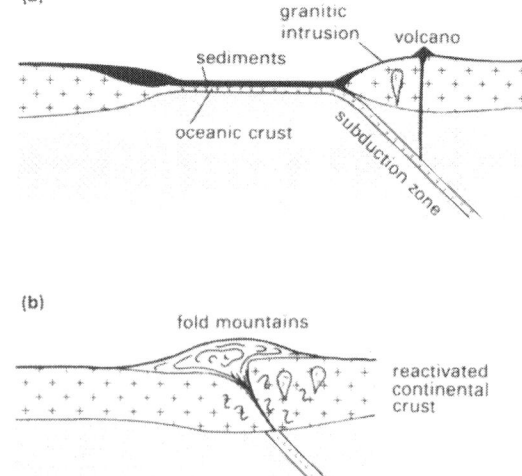

Figure 2.2 Schematic cross sections to illustrate how continental collision can lead to crustal reactivation. (a) As an ocean closes between two continental masses, sediments and volcanic rocks accumulate on the ocean floor and along the continental margins. Magmatic activity associated with subduction heats the overlying continental crust. (b) After collision, the sediments and associated igneous rocks are deformed and metamorphosed and represent an addition to the continental crust. Older crust on the subduction side of the original ocean basin is likely to experience widespread reactivation, including emplacement of granites, having already been heated by magmatic activity associated with the subduction. Crust on the other side is less likely to be reactivated.

14

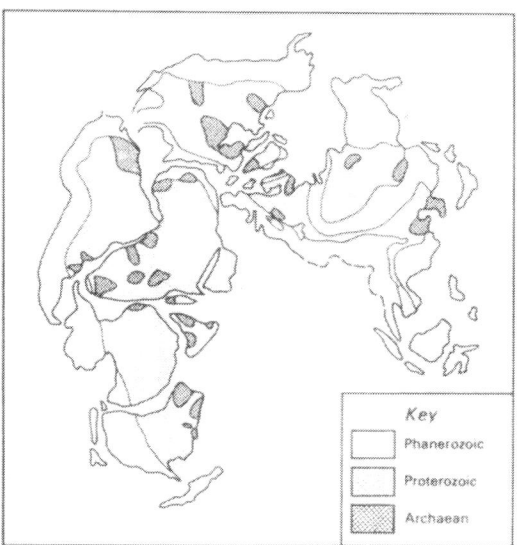

Figure 2.3 Simplified global map of age provinces on continents in their presumed pre-Mesozoic positions, to illustrate how the area of tectonically stabilised continental crust has progressively increased since the Archaean.

Key
Phanerozoic
Proterozoic
Archaean

2.2 Orogeny and thermotectonic events

Orogenic belts are typically elongate zones of crustal deformation and regional metamorphism, e.g. the Himalayan–Alpine chain. There are some obvious elongate belts on Figure 1.3: the Kibaride and Damaride belts, for example. Equally obvious are broad non-linear areas, both among the cratons and especially among the younger Pan African mobile belts.

When the enormous extent of the Pan African area was first recognised early in the 1960s, doubts were expressed about the significance of the c. 550 Ma ages that are most frequently obtained from the rocks in it. No orogenic event, as normally understood, seemed capable of producing deformation and regional metamorphism on so huge a scale. The less specific term **thermotectonic event** was coined to imply that rocks have been heated up and deformed on a regional scale, but that the processes involved were not necessarily those of a 'typical' orogeny. The term is frequently shortened simply to 'event', and events are given names, the best-known being the Pan African event. To conform with current usage, the term *event* will be used in this text, although later chapters will attempt to show that the terms event and orogeny can be regarded as synonymous – i.e. that the processes summarised in Figure 2.2 are relevant to the geological history of Africa.

2.3 Regional patterns in the Precambrian of West Africa

Figure 2.4 contains maps and diagrammatic cross sections of parts of the three main age provinces in West Africa. They show that there is a fundamental similarity in the geological relationships among Precambrian rocks throughout most of the region, irrespective of the age province in which they occur.

The three areas in Figure 2.4 have the following features in common:

(a) A basement (or **basement complex**) is dominated by mainly amphibolite-grade quartz–feldspar–biotite (±hornblende) gneisses and migmatites. They range in composition between granite and diorite, but probably average out at granodiorite. Layers and lenses of other rock types, such as quartzites, marbles and amphibolites, also occur within the basement but, although they may be widespread,

metasomatism, partial melting and granitisation, and further deformation in regions of high pressure and temperature. The crust has been reactivated.

When the crust is reactivated, pre-existing geological structures and radiometric ages are *overprinted* by later ones. The earlier record is not totally obliterated, but careful mapping, sampling and geochronological analysis are essential if that record is to be interpreted properly.

Age provinces on maps such as Figures 1.3 and 2.1 therefore record principally a history of progressive crustal stabilisation (or cratonisation) since the Archaean. Progressively larger areas of continental crust have become stabilised in successive orogenies. This important conclusion is strikingly summarised in Figure 2.3, which shows the continents in their presumed pre-Mesozoic positions, that is, their positions prior to the last major episode of continental fragmentation. It is a highly simplified map of Archaean, Proterozoic and Phanerozoic age provinces. At the end of the Archaean there was very little tectonically stable crust and large areas were still subject to reactivation by repeated orogenic events. By the end of the Proterozoic, the picture had reversed: the area of tectonically stable crust now greatly exceeded that of crust still subject to reactivation.

they are quantitatively subordinate. The complex structures and geochronological relationships among the basement rocks provide much of the evidence for crustal reactivation (Section 2.1). In other words, the basement is made up of rocks that have experienced the effects of more than one thermotectonic event.

(b) Elongate **supracrustal belts** consist of mainly greenschist to amphibolite facies phyllites, greenstones and schists. These rocks occupy synclinorial structures within the basement. They have a varied lithology and mineralogy and represent sediments and volcanics of many different types. They are called supracrustals because they look as though they were originally deposited upon the basement and are therefore younger. Their contact relationships with basement rocks are (where exposed) commonly gradational or sheared due to faulting or thrusting, and direct evidence of superposition is rare. Though tightly folded and commonly faulted, they generally have less complex structural and geochronological relationships than surrounding basement rocks, and are generally believed to have been deformed and metamorphosed during the last of the thermotectonic events which reactivated that basement.

(c) Syntectonic to late-tectonic *plutonic intrusions* are mainly of granite to granodiorite in composition, but include smaller masses of diorite, gabbro, syenite and related rocks. They intrude both basement and supracrustals. Contacts with basement rocks vary from concordant and gradational to sharp and cross cutting; contacts with supracrustals are normally sharp and cross cutting. Their field and geochronological relationships indicate that emplacement occurred during the last of the reactivation events to affect the basement – i.e. during deformation and metamorphism of the supracrustals.

Important: Some of the supracrustal and granitic rocks in the Pan African age province (of which Fig. 2.4c shows a part) may be lowermost Phanerozoic, in terms of deformation, emplacement and perhaps even deposition. But the rocks contain no fossils and they are part of a much greater region of metamorphic and intrusive rocks that are known to be Precambrian. It is thus convenient to consider the Pan African age province under the general heading of Precambrian.

The relative proportions of the three components vary somewhat from region to region. In particular, the Birimian supracrustals (Fig. 2.4b) are more widespread than those of both the Archaean (Fig. 2.4a), and the Pan African (Fig. 2.4c). Even allowing for such variations, however, the overall similarity displayed by the different parts of Figure 2.4 is striking. Moreover, throughout most of West Africa, the structural grain of the Precambrian rocks generally lies between the N–S and NE–SW trends that are shown in Figure 2.4. This grain is defined by the orientation of the supracrustal belts and by the strike of foliation in the schists and gneisses.

Similarities between the different age provinces extend to the rocks themselves as seen in outcrop. These are reviewed in the next section, partly to avoid tedious repetition later, when the individual age provinces are described and discussed separately, and partly to re-emphasize the crucial point made in Section 1.7, that *there is no relationship between the age of a rock and the degree of metamorphism or deformation that it has experienced.*

2.3.1 Basement rocks

Irrespective of their age, the basement rocks everywhere are predominantly pink to grey coloured quartzo-feldspathic gneisses and migmatites. Other rock types commonly encountered in basement terranes include quartzites (sometimes micaceous, sometimes magnetite- or haematite-bearing), quartz–mica–schists, amphibolites, marbles and calc-silicates. The relative proportions of these additional rock types vary enormously from place to place; in some regions they are relatively common, whereas in others they may be totally absent over extensive areas. Although the metamorphic grade of basement rocks is mainly in the amphibolite facies, substantial areas may be underlain by rocks metamorphosed in the granulite facies.

The distinction between gneiss and migmatite is not an easy one to make, particularly as geologists working in different parts of the basement do not always use the terms in the same way. The term *migmatite* literally means 'mixed rock', implying the presence of both metamorphic and magmatic (granitic) components, typically segregated into bands. The bands may be formed by partial melting within the rock itself, so that low-melting granitic constituents migrate and segregate into separate bands. Such rocks typically show evidence of ductile deformation – i.e. deformation in the semi-molten state – producing complex fold patterns.

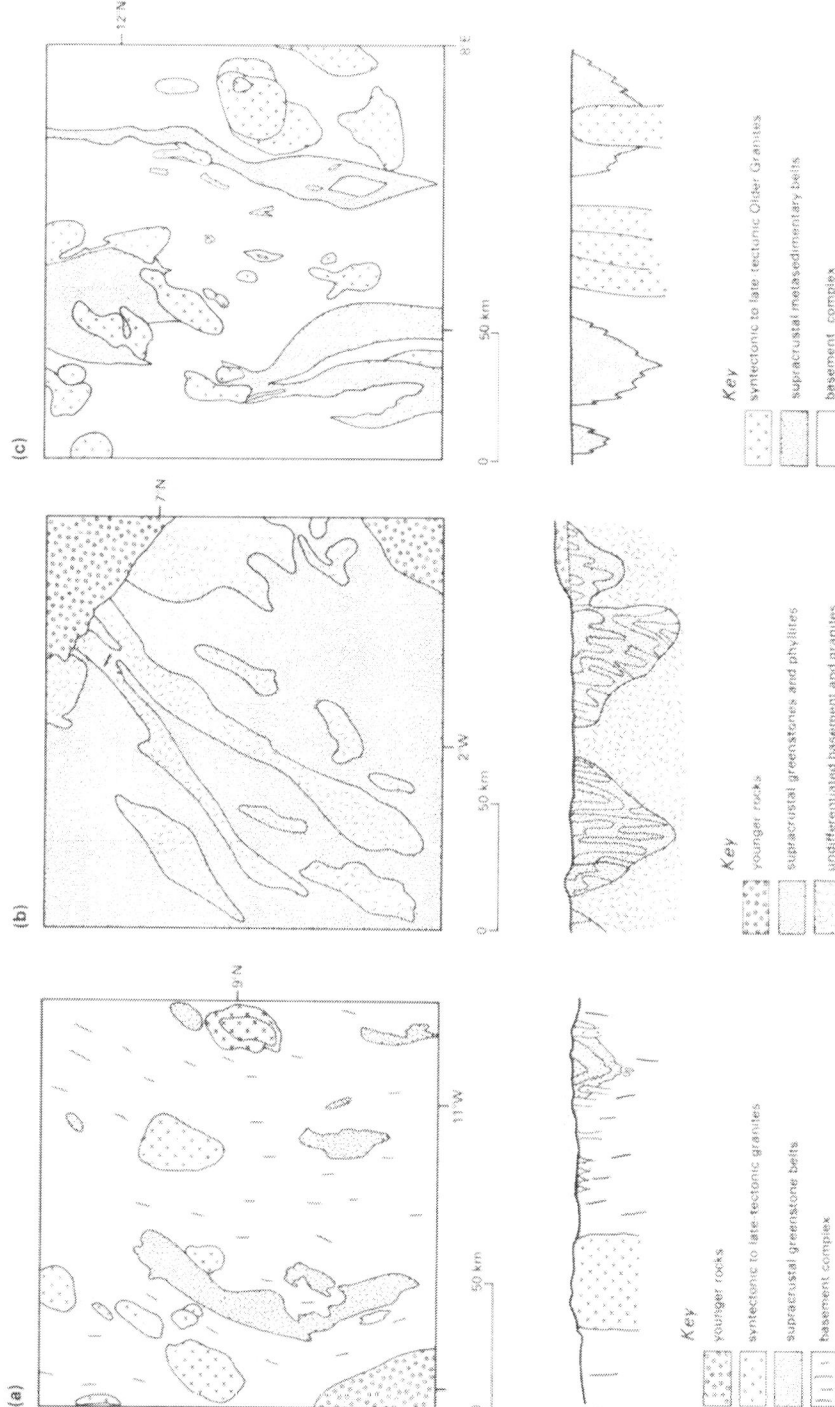

Figure 2.4 Maps and generalised and vertically exaggerated cross sections illustrating the fundamental similarity of geological relationships in the Precambrian of West Africa. See Figure 2.1 for location of each map. (a) Archaean: part of northeastern Sierra Leone and adjacent Guinea. (b) Lower to mid-Proterozoic: part of southern Ghana (c) Upper Proterozoic (Pan African): part of northern Nigeria.

Key

younger rocks

syntectonic to late tectonic granites

supracrustal greenstone belts

basement complex

Key

younger rocks

supracrustal greenstones and phyllites

undifferentiated basement and granites

Key

syntectonic to late tectonic Older Granites

supracrustal metasedimentary belts

basement complex

There is no hard-and-fast distinction between migmatites and banded gneisses in which the bands may be inherited from original sedimentary layering or from inhomogeneities in original igneous rocks (large phenocrysts, inclusions, pegmatite veins, and so on). Banding in gneisses is not necessarily parallel to original layering and may be enhanced by metasomatic processes leading to differential segregation (or metamorphic differentiation) into darker and lighter bands.

Granitic gneisses are foliated or lineated rocks of granitic to granodioritic composition which are generally weakly banded and commonly have gradational boundaries with banded gneisses and migmatites. They are probably pre-existing granitic intrusions that have been partly deformed and recrystallised during the last phase of reactivation to affect the basement. High-grade metamorphic rocks of original igneous parentage are often referred to as **orthogneisses**, those of sedimentary origin as **paragneisses**.

Where granitisation in the basement has been particularly intense, whether as a result of metasomatism or partial melting or both, migmatites and gneisses grade into intrusive granites. Where this occurs, the inhomogeneities that characterise metamorphic rocks give way to the more uniform textures of igneous rocks. In basement terranes such igneous rocks are the deeper parts of the large syntectonic granites (see Sec. 2.3.3). Veins of coarse-grained **pegmatite** and fine-grained **aplite** occur throughout the basement, but they are commoner round granitic intrusions, where they may form a high proportion of the outcrop in some places, cross cutting and engulfing blocks of country rock in complex **agmatitic** patterns.

Foliation in the basement is almost everywhere steeply dipping and trends mainly between N–S and NE–SW (Fig. 2.4). However, in detail the structures in basement terranes are extremely complex because the rocks have endured repeated deformation during a series of reactivation events, each of which involved widespread migmatisation and granitisation. The structures in the main probably record the effects of the last reactivation to have affected a particular basement area. The last reactivation in each age province was also responsible for the deformation and metamorphism of the rocks now occupying the supracrustal belts, which form the subject of the next section.

Before moving on, however, it is worth re-emphasising that the features summarised here may

be encountered in the basement areas of any part of West Africa, irrespective of its age. They are also typical of gneisses and migmatites in other parts of the world. A good deal more variation is found among the rocks of the supracrustal belts in the different age provinces.

2.3.2 Supracrustal rocks

The schists and phyllites which now make up the supracrustal belts represent an assortment of sedimentary and mainly volcanic igneous lithologies, the proportions of which differ considerably from belt to belt and in particular from region to region. Among the **metasediments**, mica–schists and phyllites generally predominate. These are rocks of **pelitic** composition, signifying their originally argillaceous nature as muds, silts and shales. Also common are **psammitic** compositions, quartzites and quartzo-feldspathic rocks, originally quartz–sandstones, arkoses and greywackes. These are arenaceous rocks, also sometimes called arenites. Conglomeratic facies may be locally abundant. In some belts, the pelitic schists and phyllites are rich in manganese oxides, and some of the quartzites are banded with magnetite or haematite (**banded iron formation**) or itabirite. Calcareous lithologies are generally less common.

Among the **metavolcanics**, greenstones, amphibolites and chlorite-rich schists are the predominant rock types, representing basaltic lavas, ashes and pyroclastics and minor intrusives, and often called metabasic rocks or simply metabasites. Amphibolites of andesitic rather than basaltic composition are plentiful in some belts, and small amounts of acid volcanics (dacites and rhyolites) are quite widespread.

The relative proportions of these two main groups of rocks differ considerably from region to region. In the Archaean age province, metasediments and metavolcanics bulk about equally; in the Proterozoic terrane, metasediments are somewhat more abundant than metavolcanics; whereas in the Pan African age province, they overwhelmingly dominate the supracrustals, with metavolcanics being greatly subordinate.

Metamorphic grades of the supracrustal belts are mainly in the greenschist to lower amphibole facies, locally reaching higher grades in the amphibolite and even granulite facies. It is sometimes possible to show that metamorphic grades in the supracrustals increase towards boundaries with basement and large syntectonic granite bodies. In some places, the metamorphic grade is low enough for original igneous and sedimentary features to be preserved, including pillow structures in lavas and graded bedding and cross stratification in sediments.

Structurally, the rocks of the supracrustal belts appear to be relatively simple, when compared with those of the basement, even though in some belts several phases of deformation can be recognised. The structures appear simple mainly because the folding is almost everywhere tight to isoclinal on all scales, with a steep axial plane foliation. The strike of the foliation is predominantly between N–S and NE–SW along the trend of the belts themselves, which are the predominant indicators of the regional structural grain (cf. Fig. 2.4). The quartzites in particular form long narrow strike ridges, but other lithologies can also have some linear topographic expression.

The supracrustal rocks represent sediments and volcanics that were deposited prior to the last reactivation of the adjacent or underlying basement, and were themselves deformed and metamorphosed during that last event. Layers and lenses of amphibolite, quartzite, marble, mica–schist and other rocks that occur among the basement gneisses and migmatites are believed mainly to represent remnants of older supracrustal sequences (see Section 2.4.1). Some are probably small outlying slices and inclusions of the main supracrustal sequences, which have been separated from the larger belts tectonically or by erosion or both.

2.3.3 Granitic intrusions

These rocks vary in dimensions from great elongate batholiths, tens of kilometres long and several kilometres across, aligned more or less parallel to the structural grain, to small subcircular stocks only a kilometre or two across. The larger bodies tend to be foliated and many of them are porphyritic, with feldspars reaching several centimetres in size, whereas the smaller intrusions are unfoliated and generally fine-grained. Most of the inselbergs that characterise many of the landforms in the West African Precambrian are formed by erosion of these granitic masses, with their characteristic exfoliation weathering patterns.

Where they are emplaced in the basement, the batholithic granites are commonly elongate and concordant and have gradational boundaries with surrounding gneisses and migmatites. The basement rocks become progressively more granitic in appearance as the intrusions are approached. For example,

pegmatitic and aplitic veining becomes more common, there is metasomatic growth of feldspar crystals and banding becomes more diffuse. The smaller granite bodies generally have sharp boundaries that cut across the basement structures. Where they are emplaced into the supracrustal belts, however, even the larger granites generally have sharp cross-cutting boundaries. Intrusions of whatever size within the supracrustals may effect contact metamorphism of the schists and phyllites, leading to the development of 'spotted' rocks, due to the growth of minerals such as cordierite and andalusite.

Features of the larger concordant batholithic granites indicate that they are syntectonic; that is to say, they were emplaced during the climax of deformation and metamorphism, when temperatures and pressures were at a maximum, and there was extensive partial melting and metasomatism in the crust. Pegmatites may be extensively developed in them. The unfoliated nature of the smaller cross-cutting (discordant) granites indicates that they were emplaced later, after the main climax of deformation and metamorphism. There are, of course, all gradations between these two extremes, and in some areas it has been possible to identify a number of stages of granite emplacement, on the basis of field relationships, textures, composition, and so on.

The granites contain variable amounts of inclusions of country rocks, particularly around their margins. In the larger batholithic granites, with their more pervasive and gradational boundaries, these inclusions have been largely 'digested', obscuring their original nature, and modifying the granite, generally making it more basic. In the smaller cross-cutting bodies, however, original textures and lithologies can often be identified – e.g. biotite-rich gneisses, amphibolites and folded metasediments – because the inclusions are rarely altered to any extent. The intrusions are not all of granitic composition. The larger batholiths range from adamellite to granodiorite, but the smaller bodies are more variable, with compositions between true granite and diorite and syenite being represented; and gabbros are also sometimes seen.*

These other intrusive rock types almost certainly represent new additions to the crust, that is to say

---

Important note: In this and subsequent chapters the terms *granite* and *granitic* are generally used in the broad sense, encompassing intrusive rocks that range in composition from quartz–diorite (tonalite) and trondjhemite (albite–granite) through granodiorite and adamellite to true granite. Another term commonly found in the literature is *granitoid*.

they are probably derived from beneath the crust, in the upper mantle. This may apply to some of the granitic intrusions also, but many of the large syntectonic batholithic masses are probably products of deep-seated metasomatism and partial melting of older rocks – i.e. they represent crustal remobilisation and recycling, rather than new additions to the continental crust.

2.4 Patterns of crustal reactivation

Previous sections have summarised many geological similarities between the different age provinces of Figure 2.1. The Precambrian crust of West Africa appears to have been subjected to a series of essentially similar cycles of events throughout most of its history. Each cycle involved deposition of supracrustal sediments and volcanics, followed by deformation and metamorphism, basement reactivation and emplacement of granitic intrusions.

The tectonic environment of each cycle must have been broadly similar throughout the region. For instance, the consistent regional trend of structures, between N–S and NE–SW (cf. Fig. 2.4), points to the existence of a fundamental and long-lived stress system. This is also indicated by many major faults and shear zones which have the same general N–S to NE–SW orientation throughout the region. They are often marked by the occurrence of mylonites and pods and lenses of vein quartz along them (although quartz veins are also found filling fractures and faults of any size and orientation). Some of these major faults involve transcurrent movements of several kilometres, and they can be identified on satellite imagery. They may represent fundamental crustal breaks and, as some of them occur along the margins of supracrustal belts, they may have controlled the development of those belts. Moreover, the fact that metamorphic and granitic rocks from one age province differ in no obvious way from those of another suggests that the pressure–temperature conditions of each reactivation event must have been broadly similar. This is also borne out by the similarity of structural styles among the deformed rocks – especially the almost ubiquitous tight to isoclinal folding in supracrustal belts.

A remarkable feature of these similarities among the different age provinces is the range of geological time involved: late Archaean through Proterozoic to early Phanerozoic. According to several authorities, the global tectonic patterns that controlled crustal development underwent significant changes at the Archaean–Proterozoic boundary, and perhaps again in the later Proterozoic (Secs 1.2 and 1.4). Yet the patterns displayed in Figure 2.4 and summarised in this chapter suggest that changes in the thermo-tectonic processes of crustal evolution since the Archaean have been changes of degree rather than of kind. That is, those processes were not fundamentally different 2500 Ma ago from what they are today, at least so far as West Africa is concerned. Indeed, cross sections that resemble those in Figure 2.4 have recently been drawn for parts of the Himalayan chain in Pakistan.

However, there is still no clear consensus about just how sedimentation and subsequent thermo-tectonic processes led to the patterns of distribution of basement, supracrustal belts and granitic intrusions portrayed in Figure 2.4. Two of the more important questions relating to the origin and development of supracrustal belts are: first, do the belts of a particular age represent originally separate basins, or were any of them once laterally continuous (Fig. 2.5); secondly, were the basins originally floored by continental crust (**ensialic** basins) or oceanic crust (**ensimatic** basins), or did both types exist, and were ensialic basins related to plate tectonics (Fig. 2.6) or some other mechanism? Answers to these questions remain inconclusive. They must depend partly on the constraints placed by palaeo-magnetic and other data upon the limits of relative movement of different segments of the African continent in the geological past (Sec. 1.4).

Where preserved, sedimentary lithologies in the supracrustal belts indicate both shallow- and deep-water deposition in different places. Supracrustal sequences containing large proportions of basic rocks are at least consistent with depositional basins floored by oceanic crust (ensimatic basins). Estimates of the total thickness of some supracrustal sequences are as high as 10–12 km, but it is not certain how much of this may be due to repetition as a result of tight to isoclinal folding.

The rarely exposed boundaries between supracrustals are usually either gradational, or faulted and sheared. Unconformities are very rare, but where found they demonstrate that some supracrustals were deposited upon continental crust. Gradational boundaries are almost as rare, but evidence of original relationships has been obliterated from them by granitisation effects in the adjacent basement which may permeate the supracrustal rocks for distances of hundreds of metres. Faulted or sheared boundaries

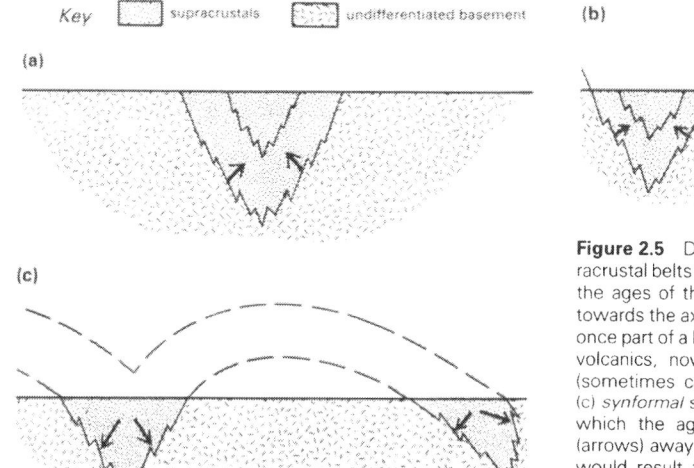

Key [ ] supracrustals [ ] undifferentiated basement

(a)

(b)

(c)

Figure 2.5 Diagrammatic cross sections showing how supracrustal belts may be: (a) isolated synclinorial structures, with the ages of the rocks becoming generally younger (arrows) towards the axial regions; (b) synclinorial structures that were once part of a laterally continuous sequence of sediments and volcanics, now separated by anticlinorial basement areas (sometimes called 'moles' by French-speaking geologists); (c) synformal structures that are really inverted anticlinoria, in which the ages of the rocks become generally younger (arrows) away from the axial region – this kind of relationship would result from refolding of great nappe-like recumbent folds.

Figure 2.6 Diagrammatic cross sections illustrating how both ensialic and ensimatic depositional basins can (a) form and (b) evolve into supracrustal belts as in a reactivated basement, as a result of plate tectonic processes followed by uplift and erosion (cf. Fig. 2.2). Ensialic basins develop by rifting upon thinned continental crust, as a result of tension produced above a subduction zone. Ensimatic basins are seas or oceans floored by basaltic crust, and may also form by more extreme tensional rifting above the subduction zone. Deformation of ensialic basins may result from a combination of factors, including differential uplift, the emplacement of granitic intrusions, or simply from compression as a result of collision following closure of ensimatic basins. Both types of supracrustal belt can have faulted boundaries with adjacent continental crust.

(a)

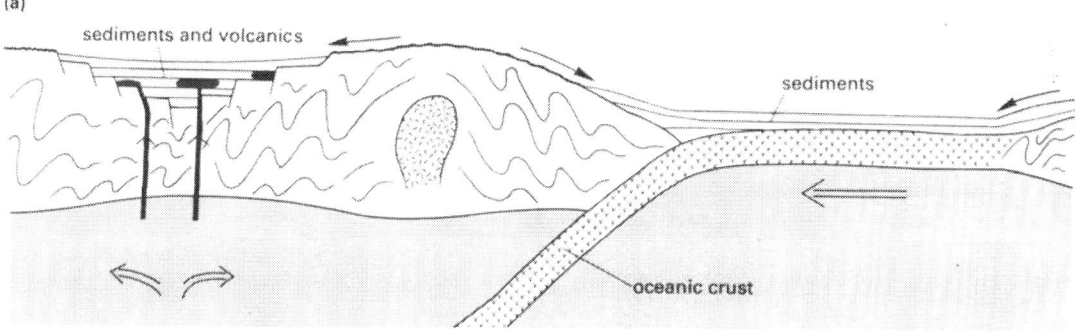

(b)

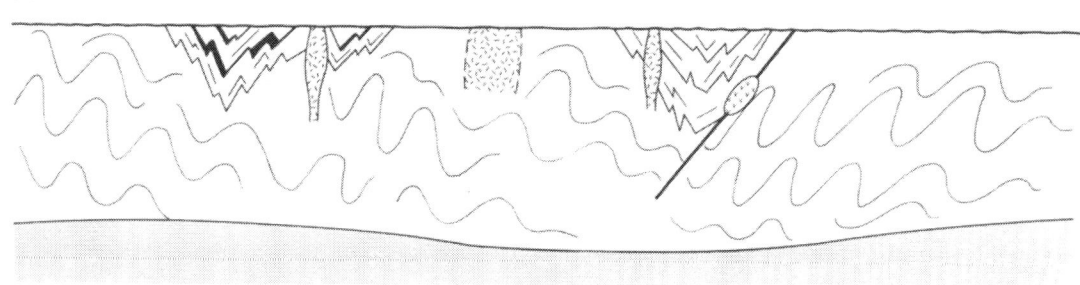

are commonest and, as suggested earlier, they may represent pre-existing zones of weakness that controlled the original basins of deposition.

2.4.1 Older supracrustal relics

As noted in Section 2.3.2, basement rocks of a particular age province may include volumetrically small amounts of quartzites, marbles, amphibolites, and so on. Many of these may represent remnants of older supracrustal sequences dismembered and engulfed in the basement during successive reactivations. Figure 2.7 illustrates very schematically how this might come about, and how pre-existing granites are turned into granitic gneisses (orthogneiss). It also shows how older supracrustal relics can be progressively eroded from the antiformal regions between the belts of downfolded younger supracrustal sequences.

Another possible reason for the progressive obliteration of these older relics lies in the composition of the rocks themselves. Most of the sediments were derived by erosion of granitic gneiss–migmatite basement terranes in the first place, and in many cases their mineralogy was dominated by assemblages of quartz, mica and detrital feldspars. Some of the volcanic rocks were of intermediate to acid composition (andesite to rhyolite). A high proportion of supracrustal sequences thus had an overall composition not very different from that of granite. Such rocks would have been susceptible to the partial melting and granitisation processes that accompanied each reactivation, and could themselves become part of the gneiss–migmatite basement, which therefore probably comprises both paragneisses and orthogneisses.

On the other hand, clay-rich sediments, quartz–sandstones and limestones are less prone to transformations in this way, and the same applies to basic volcanic rocks. It may be no coincidence that the rock types most commonly encountered among the demonstrably older supracrustal relics in basement terranes are pelitic metasediments rich in aluminosilicate minerals, quartzites, marbles and amphibolites.

The occurrence of older supracrustal relics in basement terranes provides additional evidence for crustal remobilisation. Areas of basement that have been subjected to repeated cycles involving deposition of supracrustals, regional deformation, metamorphism and reactivation, followed by erosion and renewed deposition of supracrustals, are commonly referred to as being **polycyclic**.

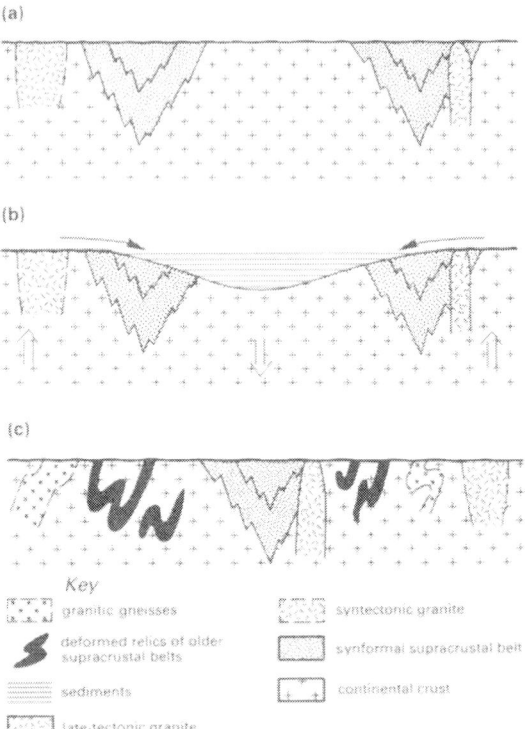

Figure 2.7 Stages in the modification of supracrustals and granitic intrusions during later reactivation of the basement. (a) Final stage of previous deformation and reactivation. For simplicity, only two synformal supracrustal belts are shown, and also two granites, one late-tectonic, the other syntectonic. (b) Formation of an elongate subsiding sedimentary basin (this may be ensialic or ensimatic), filled with volcanics and sedimentary material eroded from the marginal areas, which rise to compensate for the subsidence. (c) Final stage of subsequent deformation and reactivation, after prolonged erosion. A new synformal supracrustal belt has formed, and a new generation of both syntectonic and late-tectonic granites has been emplaced. Complexly deformed relics are all that remain of the older supracrustal belts, and pre-existing intrusions have been made over to granitic gneisses. Erosion may be a factor in the elimination of older relics, because they will tend to be in the antiformal regions between the new synformal supracrustal belts.

2.5 Summary of Precambrian events in West Africa

Table 2.1 lists the names that have been given to major rock sequences and events involved in the evolution of West African continental crust. Some of the names have appeared in earlier sections, but it

Table 2.1 Principal rock sequences and events in the Precambrian crustal evolution of West Africa (wavy lines denote major discontinuities).

Era	Supracrustal sequences	Regional metamorphic (reactivation) event
Phanerozoic	African shield stabilised except for Hercynian and Alpine effects at margins	
Proterozoic	~~~~~~~~~~~~~	*c.* 550 Ma (Pan African, including Rokelide belt in west)
	Katangan	
	?~~~~~~~?	*c.* 1100 Ma (Kibaran)
	~~	
	Birimian	*c.* 2000 Ma (Eburnian)
	~~	
Archaean		*c.* 2750 Ma (Liberiuan)
	greenstone belts	
	~~	
		c. 2950 Ma (Leonian)

is appropriate to place them all in context before embarking on descriptions of individual age provinces.

The **Liberian** was the last major event to affect the Archaean cratonic nucleus, and was responsible for metamorphism and deformation of the supracrustal **greenstone belts**. There is increasing evidence of an earlier event, the **Leonian**. Since the Liberian, the Archaean nucleus has been tectonically stable.

The large area of Proterozoic rocks east and north of the nucleus is dominated by thick supracrustal sequences of the Birimian. There is as yet no conclusive evidence for a Liberian or older basement to the Birimian supracrustals, but it seems likely that one did exist (cf. Fig. 2.4b). The Birimian supracrustals were deformed, metamorphosed and tectonically stabilised in the **Eburnian** event, and became part of the West African craton.

On both sides of the craton are areas of more recently deformed rocks. In the west, the **Rokelide belt** (Fig. 2.1) of late Proterozoic sediments and volcanics lies within a reactivated Archaean basement. In the east, the supracrustal belts have collectively been referred to as the **Katangan** but the time of their deformation and metamorphism remains uncertain. Some of them may have been affected by the **Kibaran** event at *c.* 1100 Ma ago, but most were probably deformed by the major Pan African event (*c.* 550 Ma). There is geochronological evidence that the basement in this large region east of the craton was affected by the Eburnian event, and perhaps also by the Liberian – in other words, the crust east of the craton has been subjected to four reactivation

events (Fig. 2.1). Since the Pan African, however, this whole region has become as tectonically stable as the older craton.

As already outlined in Section 1.7, the rocks of this Pan African age province east of the craton were formerly assigned to the Dahomeyan at a time when the rocks of this region were considered to be the *oldest* in West Africa. As they are now known to include the youngest supracrustals and the last basement areas to be reactivated, use of the name should be restricted to the basement complex of the Pan African terrane east of the craton; and the supracrustal belts can be called Katangan.

Bibliography

Clifford, T. N. 1970. The structural framework of Africa. In *African magmatism and tectonics*, T. N. Clifford and I. G. Gass (eds). London: Oliver & Boyd.

Coward, M. P., M. Q. Jan, D. Rex, J. Tarney, M. Thirlwall and B. F. Windley 1982. Geo-tectonic framework of the Himalaya of N. Pakistan. *J. Geol. Soc. Lond.* **139**, 299–308.

Dessauvagie, T. F. J. 1974. *Geological map of Nigeria*, 1 : 1 million. Lagos: Nigerian Min. Geol. & Met. Soc.

Kennedy, W. Q. 1964. *The structural differentiation of Africa in the Pan African*. Res. Inst. African Geology (Leeds), 8th Ann. Rep., 48–9.

Myers, J. S. 1978. Formation of banded gneisses by deformation of igneous rocks. *Precambrian Res.* **6**, 43–64.

Williams, H. R. 1978. The Archaean geology of Sierra Leone. *Precambrian Res.* **6**, 251–68.

Windley, B. F. 1977. *The evolving continents*. London: John Wiley.

Wright, J. B. and P. McCurry 1970. A reappraisal of some aspects of Precambrian shield geology. Discussion. *Geol. Soc. Am. Bull.* **81**, 3491–2.

Note: A wealth of information, relevant to virtually all chapters of this book and at the final-year undergraduate or research level, is to be found in *Afrique de l'Ouest: introduction géologique et termes stratigraphique*, edited by J. Fabre, published by Pergamon Press (1984).

3 *The Archaean of West Africa*

SUMMARY

The nucleus of the West African craton is characterised by the typical Archaean granite–greenstone association in which the 'granite' component (basement + intrusive granites) is dominant in areal terms. Metavolcanic greenstone lithologies form the lower parts of supracrustal sequences, particularly in the west, but become less important in the east where the sequences are thinner, higher metamorphic grades are typical and banded iron formations are a major component.

Most of the greenstone belts are synformal in character, with steeply dipping isoclinal folds, but low-angle folding and thrusting is developed in the Marampa Group of south-western Sierra Leone and Guinea. A belt of granulite facies rocks (the Kasila Group) forms the boundary of the craton in this region. It contains dismembered portions of large gabbro–anorthosite bodies and is separated from the gneiss–migmatite basement by a wide zone of sheared and mylonitised rocks.

Radiometric age determinations suggest that the major thermotectonic event responsible for the present distribution of rock types was the Liberian (*c.* 2750 Ma), though there is inconclusive evidence of earlier events. It is also possible that the Marampa and Kasila Groups may be younger than the bulk of the 'typical' greenstone belts.

Iron ores are the major mineral deposits presently being exploited in Archaean rocks, especially in Liberia. Bauxite and fluviatile deposits of rutile are mined in Sierra Leone. There is also gold and some low-grade chromite, but only traces of base metal mineralisation have been found.

3.1 Introduction

In common with most Archaean regions elsewhere in the world, the West African Archaean is characterised by the **granite–greenstone association**. The 'granite' component of this association comprises both the gneiss–migmatite basement and the intrusive granites as summarised in Section 2.3 and Figure 2.4; the supracrustal belts are the 'greenstone' component. The name comes from the green colour of the chlorite and hornblende in metamorphosed basic volcanic rocks. These typically predominate in the lower parts of greenstone belt sequences, the upper parts being dominated by metasediments. All the supracrustals are generally considered to belong to the last cycle of sedimentation and volcanism in the Archaean of this region and were deformed and metamorphosed in the Liberian. It is not certain that the basement is everywhere older than the supracrustals; some of the gneisses and migmatites may represent parts of the supracrustal sequence that have been granitised.

Figure 3.1 is a map of the Archaean nucleus of the West African craton. It underlies much of Guinea, the extreme west of Ivory Coast, and most of Liberia and Sierra Leone.

3.2 The basement complex

The basement is dominated by quartzo-feldspathic biotite- and hornblende-bearing migmatites and gneisses ranging in composition from diorite through tonalite to granite, with granodiorite predominating. There are also small lenses and sheets of amphibolite, the metamorphosed and deformed relics of basaltic dykes. The overall northerly to northeasterly structural trend in basement rocks swings round to the north-west towards the northern limits of the Archaean terrain, and locally in the south, near the Kasila belt (Fig. 3.1). Structures with an overall east–west trend in parts of Sierra Leone may represent an earlier deformation. Amphibolite facies metamorphism is characteristic of much of the basement.

The Archaean basement rocks are not formally named in Sierra Leone and Liberia, but in western Ivory Coast the main basement area is referred to as the **Migmatite and Gneiss Formation**. There are two higher-grade (granulite facies) areas known as the **Mt. Douan Formation** and the **Man charnockite complex**, which are probably older (Sec. 3.6) and may contain ancient supracrustal relics. In Guinea, there is a three-fold subdivision, into **Mahana Gneiss** east of the Simandou belt, **Macenta Gneiss** to the west, and, further to the north-west, the **Guinea Gneiss**.

25

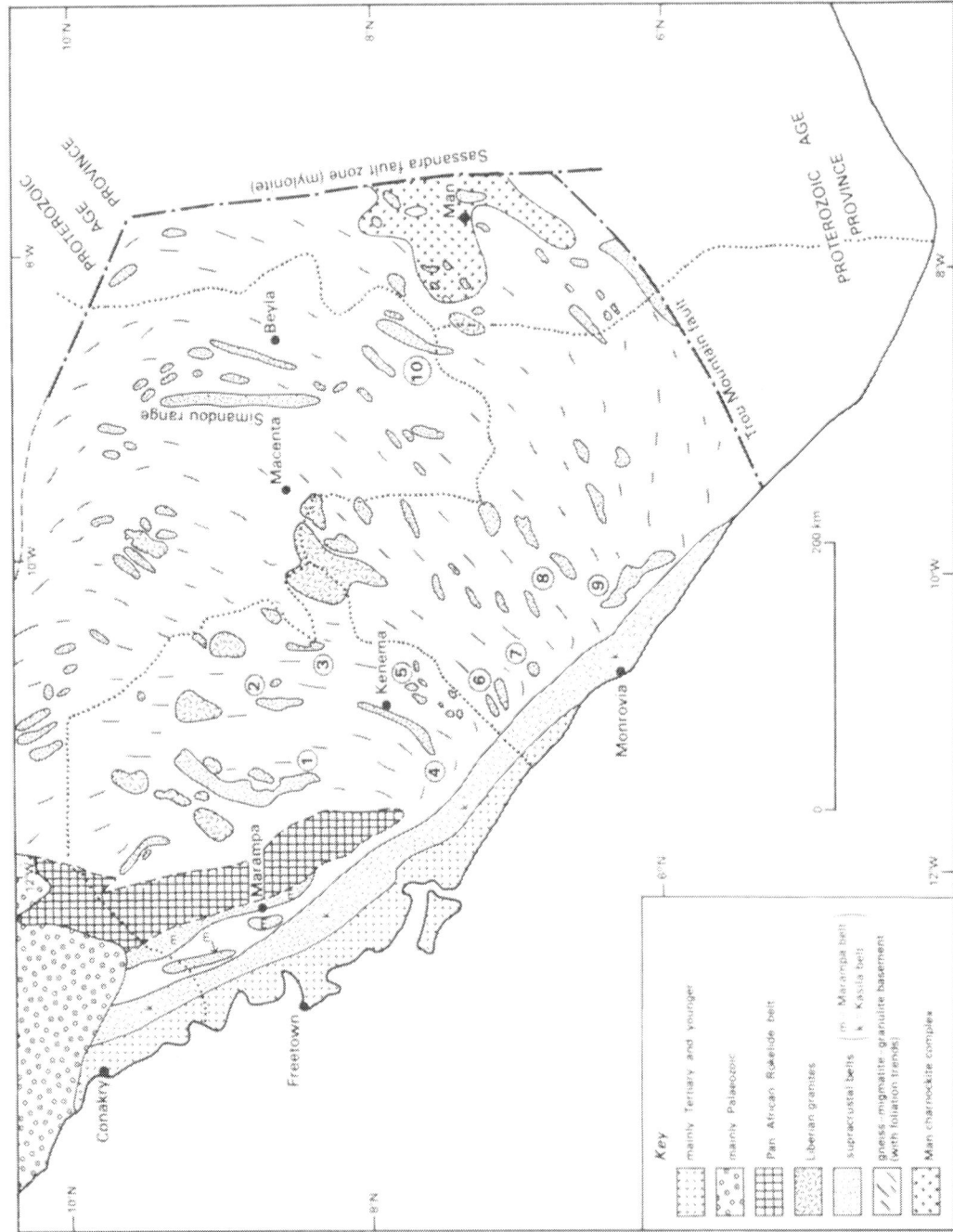

Figure 3.1 Generalised map of the Archaean age province of West Africa. Numbers refer to supracrustal belts named in Figure 3.2. Names of other belts are not shown, nor are all the granitic intrusions. The northern boundary of the age province is tentative.

3.3 The supracrustal belts

Synformal sequences of greenstones and metasediments occurring as linear belts and small relics have been mapped over a large area (Fig. 3.1). Some of the smaller relics could be remnants from an earlier sedimentary cycle, but the available geochronological evidence suggests all the supracrustals to be of the same age.

In the western part of Figure 3.1, that is over most of Sierra Leone, the belts are up to 130 km long, with successions up to 6.5 km thick, mostly metamorphosed to greenschist grade. In these belts the two-fold subdivision that characterises Archaean greenstone belts worldwide can be recognised.

The lower part of the sequence often commences with ultramafic rocks, passing up to mafic types. The ultramafic rocks were originally lavas and sills, now metamorphosed to serpentinites and chloritic schists with talc, tremolite and anthophyllite. They are interbedded with amphibolite and occasional metasediments, sometimes fuchsite-bearing (**fuchsite** is a Cr-rich mica). The overlying mafic rocks are mainly tholeiitic amphibolites with pillow structures and amygdaloidal and vesicular textures preserved, and represent pillowed and massive lavas and intrusive sills.

The upper part of the sequence is metasedimentary: quartzites (sometimes with fuchsite), micaschists, metacherts, metaconglomerates, cordierite–garnet–schists, quartzo-feldspathic schists, and banded iron formations are all represented. There are also weakly metamorphosed greywacke–turbidites with original sedimentary features such as graded bedding and flame structures preserved. A subordinate metavolcanic component includes some siliceous varieties, notably dacites and rhyolites. Granitic gneiss pebbles in the conglomeratic facies of at least some of these belts (e.g. Gori Hills, Sula Mountains) indicate the proximity of continental crust during deposition of the supracrustals.

In these greenstone belts most of the metamorphism is in the greenschist to epidote–amphibolite facies, though almandine–amphibolite facies rocks occur locally, especially near contacts with intrusive granites.

There are rapid and irregular changes in lithology and thickness within the greenstone belts, and meaningful determination of stratigraphic thicknesses is impossible, because the extent of duplication and omission due to folding and faulting is unknown. Nonetheless, thicknesses of several kilometres are common in the large belts, though in the smaller relics they may be no more than a few hundred metres.

Further to the east, in southeastern Sierra Leone, Liberia and Ivory Coast, the belts are generally smaller, not exceeding 40 km in length and rarely more than a kilometre or so thick, with metamorphism mostly in the granulite facies. These smaller belts are dominated by metasediments, particularly banded iron formation rocks, along with quartzites, mica–quartzites, metagreywackes, metacherts and pelitic schists. The subordinate mafic and ultramafic rocks are normally low in the sequence, as in the larger greenstone belts further west, and they are mainly metavolcanic amphibolites.

In western Ivory Coast, the Man charnockite complex is an extensive area of hypersthene-bearing charnockitic rocks which includes norites and anorthosites. Gabbro–anorthosites also intrude banded iron formation rocks in the Mano River area of southeastern Sierra Leone and western Liberia.

High-pressure mineral assemblages are rare in the granite–greenstone terrane as a whole. Sillimanite, staurolite, garnet and cordierite are the typical high-grade minerals, even among the granulite facies rocks, though kyanite has been recorded. These are assemblages typical of a geothermal gradient found in regions dominated by copious granitic intrusions (Sec. 3.4).

Deformation in the supracrustal belts is everywhere of a similar structural style, irrespective of differences in size, lithology and metamorphic grade. The rocks have been folded into tight anticlines and synclines, though deformation was probably more intense in the high-grade belts, where original sedimentary features are not preserved. Deformation has obscured original relationships between basement and supracrustals. There is evidence of a discordance in the Mano River area, which may mean that the supracrustals were deposited on basement in this area at least. Figure 3.1 shows the long Simandou belt in Guinea to be discordant with the regional foliation, though locally the rocks are seen to be concordant with the adjacent granitic basement. This may be a local effect, however, tectonically induced, and the regional discordance could be an unconformity between sediments and basement.

The main lithological and stratigraphical differences among the Archaean supracrustal belts of West Africa are summarised in Figure 3.2. In brief,

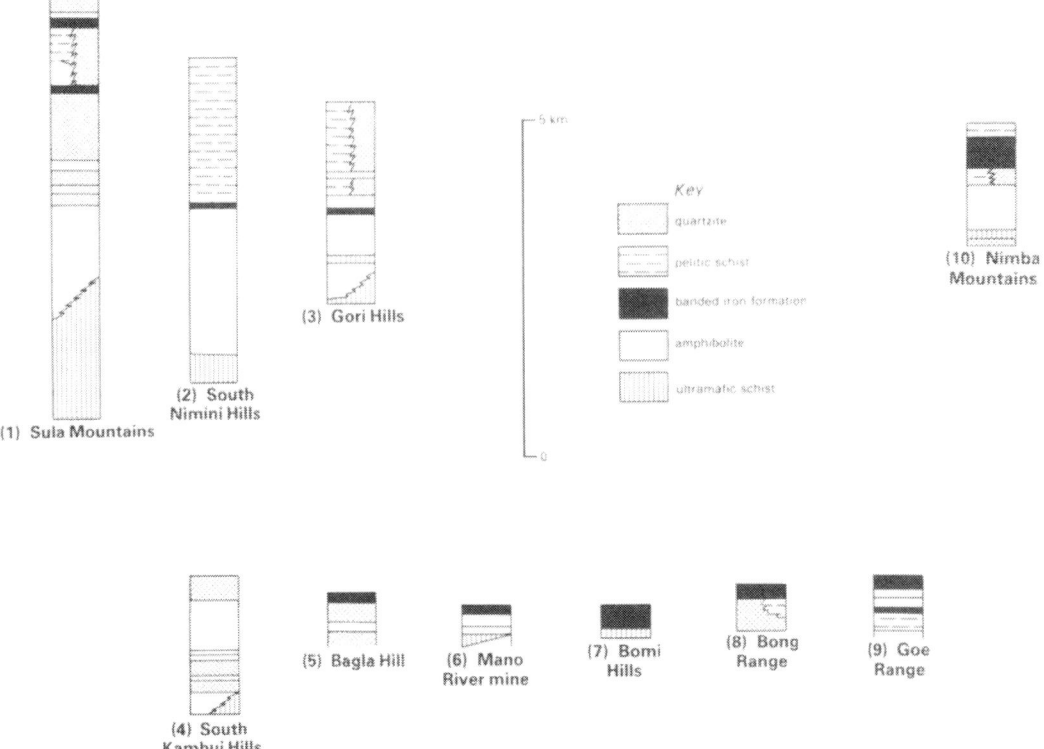

Figure 3.2 Comparative stratigraphic columns for some supracrustal belts on the Archaean cratonic nucleus of West Africa, showing how greenstone-type rocks dominate thicker sequences in the west, whereas banded iron formation rocks dominate thinner sequences in the east. The sequences are generalised, because of lateral facies changes and complications due to folding. Numbers at the base of each column refer to locations indicated on Figure 3.1, and the columns are shown in their approximate relative positions.

the belts become shorter and their successions thinner, when traced eastwards and southwards, and metamorphic grades become higher. Their lithologies also change. Greenstones become less abundant, banded iron formation rocks figure more prominently. These and other features suggest that the low-grade supracrustal belts in the west may have formed originally in deeper water and less stable tectonic conditions, perhaps in one or more ensimatic basins, while the mainly sedimentary successions in the east developed in shallower water and a more stable setting, perhaps in ensialic basins. Here chemically precipitated iron-rich sediments could accumulate during periods when the supply of clastic material was low. Mineral equilibrium studies indicate that the $P-T$ conditions of metamorphic maxima in the lower-grade greenstone belts in the west were around 600°C and 5 kbar. Higher temperatures ($c.$ 750°C) and pressures ($c.$ 7 kbar) prevailed

in the east, where granulite facies metamorphism occurred.

The supracrustal sequences have been given different names in the different countries in which they occur, although they probably all belong to the same sedimentary–volcanic cycle. The greenstone-dominated belts in Sierra Leone are called the **Sula Group**, whereas in Liberia the belts rich in banded iron formation rocks are referred to as the **Nimba Group**. Quartzites figure prominently in the supracrustals of Guinea, where they are known as the **Simandou Group** (other spellings are Siamandou and Sinandou), subordinate mafic lithologies (amphibolites and related rocks) being placed in the **Beyla Group**. In western Ivory Coast, the **Mt Gao Formation** comprises quartzites and banded iron formation, along with amphibolites and pyroxenites, described as forming tight synclines within the gneiss–migmatite basement.

28

3.3.1 The Marampa and Kasila belts

West of the main group of greenstone belts in Sierra Leone and Guinea is a discontinuous northwesterly trending strip of low-grade metamorphic rocks, called the **Marampa Group** in Sierra Leone, which has a greenstone belt stratigraphy. The lower part consists of metamorphosed basaltic and andesitic volcanics, some with pillows and porphyritic and flow textures preserved, and ultramafic intercalations, largely serpentinised. The upper part is metasedimentary, including quartzites with pebbly horizons and cross bedding, also fuchsite–quartzite and manganiferous quartzite. Other rocks include quartz–schists, mica–schists, garnet–schists and some gneisses, as well as banded iron formation. Despite some lithological and stratigraphic affinities with the synformal greenstone belts, these rocks are to be regarded as distinct from them because of their marked structural differences. They have recumbent fold structures and they are in low-angle thrust-faulted contact with the underlying granitic basement, in striking contrast to the relationships portrayed in Figure 2.4a. Metamorphic grades increase from greenschists in the east to amphibolite in the west, where there are minor granitic intrusions into the paragneisses.

A belt of granulite facies rocks lies immediately west of the Marampa Group and parallel to it. Called the **Kasila Group** in Sierra Leone, it extends northwest into Guinea and south-east into Liberia as far as 5°N. Most of this group consists of high-grade metabasic igneous rocks, now pyroxene-bearing basic granulites. Within them are lenses of layered anorthositic and gabbroic rocks, up to 200 m thick and 50 km long, which are metamorphosed and disrupted remnants of large basic intrusions. There appears to be a decrease in metamorphic grade towards the western part of the belt, where amphibolites become increasingly important. Minor quartz–magnetite, quartz–diopside and aluminosilicate-bearing rocks in the Kasila Group represent highly metamorphosed equivalents of banded iron formation, marbles and pelites respectively.

Very high pressure–temperature conditions are preserved by the metamorphic minerals of the Kasila Group. Abundant pyroxene, kyanite and sillimanite are found in Sierra Leone and Liberia, suggesting that they represent a different thermotectonic regime (possibly a deeper crustal level) from that of the granite–greenstone terrane, where high-pressure minerals assemblages are rare.

There is an overall similarity in lithological and stratigraphic character with the greenstone belts, but the grade of metamorphism is much higher in the Kasila Group, there are more layered anorthositic rocks (as is usual in deeper crustal layers) and the rocks are structurally different from those of the greenstone belts. There are relatively few minor structures, and the predominantly NW–SE trending foliation has moderate dips to the south-west throughout the belt. In addition, over much of its length the northeastern boundary of the granulites is marked by a broad zone, up to 5 km thick, of sheared gneisses and mylonites, dipping at moderate angles to the south-west and indicative of medium-angle thrusting on to the granitic basement. The western limit of the belt is not seen, as it is obscured by post-Palaeozoic sediments. Geophysical investigations in Liberia have revealed a large NW–SE trending positive gravity anomaly (c. 50 mgal) along the coast and beneath the continental shelf. It is interpreted to represent the Kasila belt, and its landward limit coincides with the mylonitic shear zone that defines the northeastern geological limit of the belt.

Evidence is presented in Section 3.6 for regarding the Marampa and Kasila belts as Archaean, but it must be recorded that many workers believe them to belong to the much younger Pan African event.

3.4 Granites

The syntectonic to late-tectonic granitic intrusions range in size from large masses having gradational contacts with the basement, through smaller cross-cutting stocks and plutons, down to pegmatites and aplites. Only the larger masses are shown on Figure 3.1. The intrusions are generally unfoliated and relatively homogeneous, and most of them appear to have a true granitic composition, though dioritic and granodioritic masses are recorded in western Ivory Coast. Hydridisation of granite with mafic members of the schist belts has occurred in places, however, and in these places compositions range from hornblende–syenite to diorite.

3.5 Fractures, faults and mylonites

Throughout the granite–greenstone terrane there is extensive development of fracture zones and shear belts, often containing mylonites. They are parallel with the generally north–south foliation trends in the granitic basement terrane, and show up on satellite photographs. Dextral displacements are recorded from many of these faults, with complementary sinistral displacements along an ENE–WSW fracture set. These fractures may represent a conjugate set consistent with a SW–NE directed principal stress. This is in accordance with thrusting movements that must have been associated with development of the mylonites and sheared rocks along the eastern boundary of the Kasila Group granulite belt. Major faults and the **Sassandra mylonite zone** help to define the eastern boundary of the cratonic nucleus (Fig. 3.1).

3.6   Correlation and geochronology

Table 3.1 provides a correlation of Archaean rocks from different parts of the West African cratonic nucleus. Numerous age determinations have been made, using a variety of methods including Rb/Sr, K/Ar and Pb/Pb measurements on both whole rocks and minerals. Not all of the age determinations are controlled by reliable geological mapping, but taken together they suggest that the supracrustal volcanics and sediments were deposited in basins developed upon or marginal to a Leonian basement and were deformed and metamorphosed in the Liberian thermotectonic event.

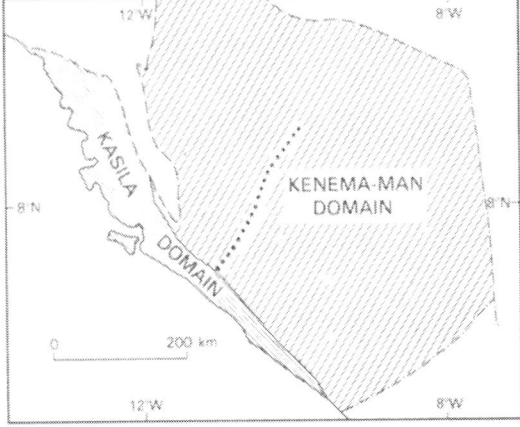

Figure 3.3 The two domains of the Archaean cratonic nucleus of West Africa. The dotted line approximately bisecting the Kenema–Man domain indicates the boundary between generally higher metamorphic grades in the east and lower grades in the west.

At this point it is appropriate to introduce the term **domain**, which is broadly synonymous with age province, but carries implications about structural and lithological relationships as well. An age province can be subdivided into two or more domains. Figure 3.3 shows two principal domains for the West African Archaean cratonic nucleus.

Table 3.1   Correlation for the Archaean of West Africa.*

Event and approximate age	Sierra Leone		Liberia	Guinea	Ivory Coast
Liberian, 2.75 Ga	deformation and metamorphism of supracrustals, basement reactivation and granite intrusion				
Broadly contemporaneous deposition of supracrustal sequences	Kasila Group (U.)	Rotokolon Fm. (metased.)	Nimba Group	Simandou and Beyla Groups	Mt Gao Formation
	Marampa Group (L.)	Matoto Fm. (metavolc.)			
	(U.)	Tonkolili Fm. (metased)			
	Sula Group (L.)	Sonfon Fm. (metavolc.)			
unconformity Leonian, 2.95 Ga	gneiss–migmatite–granulite basement			Macenta Gneiss, Mahana Gneiss and Guinea Gneiss	Migmatite and Gneiss Formation
pre-Leonian, 3.1 Ga and older	not identified but probably present				Mt Douan Formation and Man charnockite complex

* Not all the formations are discussed in the text.

The larger **Kenema–Man domain** forms most of the cratonic nucleus. It contains the mainly north to northeasterly trending supracrustal belts, which are dominated by greenstones in the west and by banded iron formation in the east. Liberian ages (*c.* 2.75 Ga) have been obtained from both basement gneisses and granites, and some gneisses have yielded Leonian ages (*c.* 2.95 Ga), implying that parts of the basement escaped the full effects of the Liberian reactivation.

The generally E–W basement structures found in parts of Sierra Leone have been ascribed to an early deformation (Sec. 3.2) and correlated with the Leonian event. However, there are areas without such structures that also give Leonian ages and the correlation can only be considered tentative at present.

A pre-Leonian age (*c.* 3.1 Ga) has been recorded from rocks of the high-grade Man charnockite complex near the eastern border of the domain, suggesting that formation of Archaean crust may have been in progress at least 3.2 Ga ago. Rocks in this region retain also the subsequent imprint of both the Leonian and Liberian events. The eastern boundary of the Kenema–Man domain is marked by major faults, and rocks in this border zone (notably in SW Ivory Coast) have also yielded ages in the 2600–2000 Ma range. They are interpreted as the results of marginal reactivation during the subsequent Eburnian event, which partly or wholly reset earlier ages. Similar relationships probably occur further north, but fewer age determinations are available and the northern limit of the Archaean nucleus shown in Figures 3.1 and 3.3 cannot be regarded as definitive. Other maps place it somewhat further north. There is some evidence of still later thermal reworking of Archaean rocks in Liberia and Ivory Coast at around 1.8–1.5 Ga, with local recrystallisation.

Rocks of the **Kasila domain** differ from those of the Kenema–Man domain in terms of metamorphic grade and structure, and their orientation is west of north rather than east of north. Some rocks from the Kasila Group have given Liberian Rb/Sr ages, but both Pan African (*c.* 550 Ma) and Eburnian (*c.* 2000 Ma) ages have been obtained from other rocks (Rb/Sr) and from minerals (K/Ar). A Liberian age has also been obtained for the Marampa Group. While it is possible to ignore the older ages and suggest that the rocks of the Kasila belt are younger than those in the Kenema–Man domain, it seems more reasonable to regard the younger ages as resulting from reactivation during first the Eburnian event (Ch. 4) and then the Pan African event, which was responsible

for deformation and metamorphism of the Rokelide belt in Sierra Leone and Guinea (Fig. 1.4 & Ch. 5). This would also be consistent with intermediate ages recorded from the basement that lies between the Kasila and Marampa belts. However, it must be re-emphasised that the Kasila and Marampa Groups are regarded by many workers as being of Pan African age.

3.7 Plate tectonics and the Liberian event

The greenstone-dominated supracrustal successions may have accumulated in ensimatic basins, the metasediment-dominated belts in an ensialic setting. It is not known whether the individual belts represent several small basins or a few large ones (Fig. 2.5), though the latter seems more likely. Assuming that plate tectonic mechanisms were involved in the evolution and deformation of these belts, then they could have been of the kind summarised in Figure 2.6. Whether ensimatic or ensialic, the basins could have been of **back-arc** or **inter-arc** type, resembling those found today above subduction zones in the western Pacific and along the South American Andes belt; though the rate and dimensions of the plate tectonic mechanisms must have been different from what they are now. Bearing these qualifications in mind, the sediments and volcanics of the Kenema–Man domain could have been deformed and infolded into the reactivated granitic basement in two main ways: either through crustal shortening during the subduction process itself, much in the way that deformation has occurred in parts of the Andean belt in South America; or through numerous collisions of small continental fragments and island arcs such as those that characterise the present-day western Pacific; or even through a combination of both processes. Whatever the mechanism, the result was formation of the tightly folded and generally north to northeasterly trending belts that lie parallel to the regional structural grain and indeed help to define it. Although every part of the Archaean domain was involved in the Liberian event, some areas were less profoundly affected than others, and it is these that have retained evidence of possible earlier structures and pre-Liberian radiometric ages.

The low- to medium-angle thrust-faulted contacts between supracrustals and basement in the Kasila domain are also consistent with a plate tectonic interpretation for the Liberian event – a continental collision of Alpine–Himalayan type. The low grade of metamorphism and large recumbent structures in the Marampa Group rocks, as well as their lower thrust contacts with the basement, suggest that they may be nappes derived from the Kasila belt. Orogenic activity within the Kasila Group prior to its high-grade metamorphism allowed the separating and sliding of high-level materials towards the north-east as nappes. Where they came to rest on the adjacent basement that was undergoing reactivation, the rocks were intruded by minor amounts of granite and metamorphosed in the greenschist facies.

The granulites of the Kasila Group would thus represent the root zones of these nappes, formed as the result of collision and suturing of two Archaean continental plates, one the West African cratonic

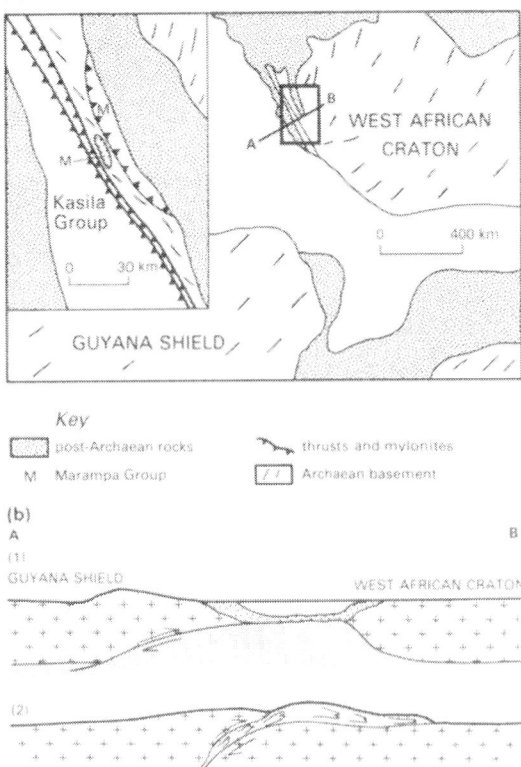

(a)

Key

▢ post-Archaean rocks   ⌁ thrusts and mylonites

M   Marampa Group   ▢ Archaean basement

(b)

nucleus, the other the Guyana shield, an Archaean nucleus in northeastern South America (Fig. 3.4). The Kasila and Marampa Group rocks are supra-crustals that originally accumulated in a basin between these two plates. The thick mylonites beneath the Kasila Group developed as these two continental masses became sutured together. The tendency for the regional grain in the basement to swing towards the north-west as the Kasila belt is approached (Fig. 3.1) suggests that there was a dextral shear component during the deformation and reactivation, that is, the collision (Fig. 3.4) was oblique rather than 'head on'.

Once again, however, it must be stressed that, while the main features of this interpretation of the Kasila domain may be valid, there are many (including probably most geologists in Liberia) who would place it in the Pan African rather than in the Archaean.

3.8 Repeating patterns of crustal reactivation – a reminder

The Archaean rocks described in this chapter under-lie the largest area in West Africa to have remained undisturbed by subsequent tectonism. However, as explained in Section 2.1, relict Archaean ages have been recorded from a number of separate localities throughout the regional metamorphic terranes out-side the area of the cratonic nucleus. This has been interpreted to mean that much of what is now the continental crust of Africa was already in existence at the end of Archaean times, *c.* 2.5 Ga ago. It was similar to that preserved on a large scale only within the area of Figure 3.1. In other words, the succes-sively younger metamorphic age provinces on maps such as Figures 1.4 and 2.1 *do not represent regions of significant formation of new continental crust in the Lower and Upper Proterozoic.* They represent regions of successive reactivation of older crust, with only comparatively minor addition of new material.

Figure 3.4 An interpretation of the Kasila domain of the West African cratonic nucleus in terms of plate tectonics. (a) Pre-Mesozoic positions of the West African craton and Guyana shield. Trend lines indicate regional structures. (b) Schematic cross-section illustrating the sequence of events leading to plate collision: (1) convergence of plates, closing of sedimen-tary basin floored by oceanic crust; (2) collision causing plastic deformation of the Kasila Group, to produce nappe sequences at high level, granulite facies shear zones at depth; (3) brittle thrusting, mylonite formation and erosion with isolation of Marampa Group as outliers (klippen).

Thus, following the Liberian thermotectonic event, the Archaean crust, consisting of gneiss–migmatite–granulite basement, supracrustal belts and granitic intrusions, acted as basement to the Birimian supracrustal sediments and volcanics, east of the cratonic nucleus. This younger supracrustal sequence was itself deformed and metamorphosed in the Eburnian event about 2.0 Ga ago (Table 2.1), so completing the tectonic stabilisation of the region now called the West African craton (Fig. 1.4). Sediments subsequently deposited upon it have remained virtually undeformed and unmetamorphosed.

The eastern part of this Lower Proterozoic crustal assemblage provided the basement for yet another set of supracrustals, the Middle to Upper Proterozoic Katangan sediments and volcanics; and it was itself further reactivated during the Kibaran and Pan African thermotectonic events, respectively about 1100 and 550 Ma ago (Table 2.1).

In summary, then:

(a) most of the continental crust in West Africa was in existence about 3000 Ma ago, and perhaps earlier;

(b) this crust, along with successively younger supracrustal sequences, has become progressively stabilised from the west;

(c) since Archaean times, reactivation has been minimal in the area of Figure 3.1, and it has been greatest in the eastern parts of Figure 1.4.

This pattern of progressive crustal stabilisation has probably been a fundamental factor in determining the occurrence and distribution of economic mineral deposits.

3.9 Economic potential in the Archaean rocks of West Africa

The Archaean province of West Africa carries mineralisation characteristic of granite–greenstone associations elsewhere. It is mainly associated with the supracrustal belts (Fig. 3.5), but these are less extensive than in the Archaean terranes of Canada, Western Australia and southern Africa, for example. Thus, the diversity of the mineral potential is correspondingly smaller. In addition, the relative scarcity of andesitic–rhyolitic volcanics among the supracrustals has limited the possibilities for finding significant base metal mineralisation of the kind associated with such rocks (e.g. Cu–Pb–Zn, As–Bi).

Generally higher metamorphic grades in the surrounding basement, coupled with its polycyclic history, mean that these rocks are generally barren of important mineralisation. Potential ore-forming elements that may originally have been present were removed by the passage of volatiles during successive reactivations, deposited at higher levels and subsequently stripped off by erosion. Mineralisation associated with the granitic intrusives is therefore most likely to be found where these are emplaced in the vicinity of the supracrustal belts.

3.9.1 Gold

Most of the documented occurrences of gold in the Archaean of West Africa appear to be in the western greenstone belts. Some 340 000 oz (c. 10 000 kg) were taken in Sierra Leone between 1930 and 1956, mainly from alluvial workings in and around the greenstone belts in the northern half of the country. Alluvial mining is still carried out by small groups of miners working river gravels, but the major commercial interest has shifted to primary sources, which have been found by tracing the alluvial deposits back to their origin. Gold–quartz veins, sometimes with pyrite and arsenopyrite or tourmaline (schorl rock), occur in amphibolites and ferruginous schists (metamorphosed banded iron formation). Some of these have been explored by drilling programmes, notably at Baomahun in the southern Kangari Hills, but also in the Nimini Hills and near Bo. Favourable grades in the range of about 10–30 ppm gold have been found. There has also been some geochemical prospecting, using arsenic and molybdenum as tracers. About 120 kg of gold were exported from Sierra Leone in 1981, but only just over 60 kg in 1983.

Further east, gold extracted from similar deposits in Liberia was worth about $3 million in 1980, and official estimates placed production at up to 45 kg per month in 1983. Exploration continues in the eastern parts of that country, where the metal is known to occur among rocks reactivated in the Eburnian event (Ch. 4).

As supracrustal belts similar to those in Sierra Leone and Liberia occur over substantial areas of Archaean terrane further north in Guinea, there must be economically workable gold deposits there also.

3.9.2 Iron ores

Iron ores form perhaps the most important mineral resource of the Archaean in West Africa. The major

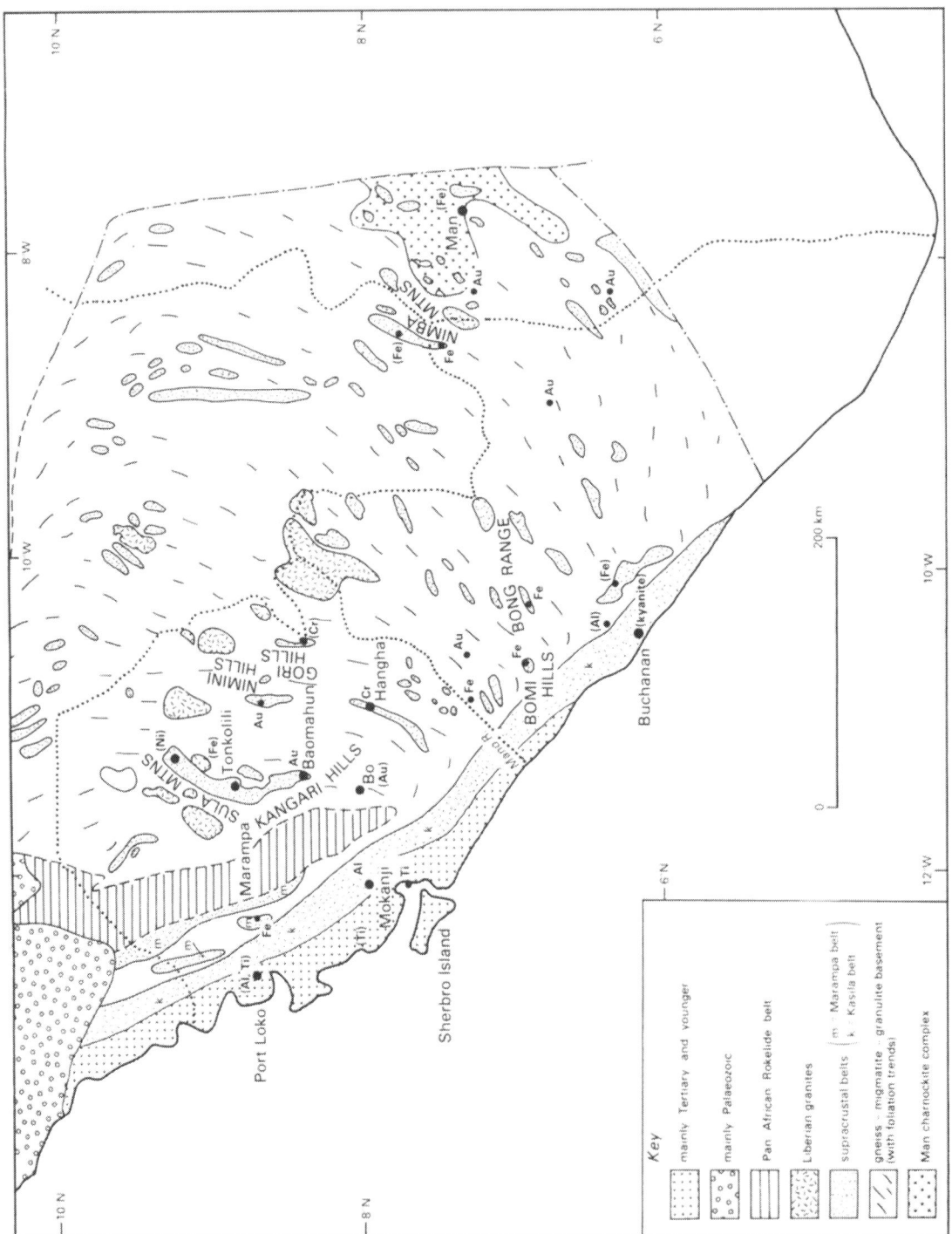

Figure 3.5 The distribution of some mineral deposits in the Archaean of West Africa. Symbols in brackets denote deposits that are known or suspected, but have not been exploited. Elements are indicated by their chemical symbols.

deposits are in the eastern supracrustal belts, which are dominated by banded iron formation. They are finely laminated rocks consisting mainly of quartz and iron oxides, mostly magnetite, though haematite also occurs, and they are interbedded with mafic schists and amphibolites, quartzites and phyllites. Liberia is a major world producer, which exported some 20 million tonnes annually, from four main centres, up to about 1980, but exports fell to 18 million tonnes in 1982 and to 15 million in 1983. The largest mine is in the Nimba belt, where the iron formation reaches 500 m in total thickness (Fig. 3.2), the others being in the Bomi Hills, the Bong Range and near the Mano River, which forms the border with Sierra Leone. Ore grades range from 35 to 70% Fe, depending on the extent of secondary lateritic enrichment, with average high-grade ore running at approximately 65% Fe. Important deposits occur in other parts in Liberia, and total reserves are believed to exceed 1 billion tonnes (10^9 t). The Nimba belt extends across the border into Guinea, where large-scale mining of the same rock units is planned, as reserves there are estimated to total 2000 million tonnes. Production should commence during the 1980s, provided that economic conditions are favourable and transportation links can be established. There are a further 7000 million tonnes estimated to be available in the Simandou belt, and deposits are also being sought in other supracrustal belts of the Guinea sector. The ore is of high quality and there are plans to use it for Nigeria's steel industry (Sec. 6.5). Substantial deposits of iron ore also occur in western Ivory Coast, in the vicinity of Man. The Mt Klahoyo deposit is reported to contain over 600 million tonnes, averaging 36% Fe, the highest grades reaching 69% Fe.

Banded iron formation rocks also occur in the upper parts of greenstone belts further west in the Kenema–Man domain, but they are generally thinner and the only major deposit so far located is near Tonkolili in the Sula–Kangari belt (Fig. 3.5). Here the banded iron formation rocks are interbedded with amphibolites and there has been considerable secondary enrichment by lateritisation processes throughout most of the Tertiary. Reserves are estimated to be of the order of 100 million tonnes, at grades of over 55% Fe. Despite extensive investigation of these deposits between 1930 and 1960, they have not been mined, partly because of their distance from suitable port facilities.

In the Kasila domain, however, the Marampa Group iron ores, haematite-rich schists among the metasediments, were successfully mined for many years, producing as much as 2.5 million tonnes annually until 1975, when the production ceased due to adverse market conditions and production problems. There are plans to recommence exploitation of these deposits during the 1980s, reserves being estimated at over 13 million tonnes of average 39% Fe ore.

3.9.3 Bauxite

Rocks of the Kasila belt contain alumina-rich facies that have given rise to secondary residual accumulations of bauxite. Although these developed as a result of tropical lateritic weathering of aluminous metasediments and anorthositic rocks during the Tertiary, the parent rocks are probably Archaean and they are considered here. Bauxite ores are formed by intense surface leaching of alkaline and basic elements and silica under conditions of high temperature and rainfall, producing initially aluminosilicate clay minerals, which are transformed by further leaching into hydrated aluminium oxides, mainly gibbsite. High-grade bauxites should contain between 40 and 60% Al_2O_3 and ideally should not contain more than 25% Fe_2O_3.

The major deposit is in the Mokanji Hills of Sierra Leone, discovered in 1960 and mined since 1963. Production was more than 800 000 tonnes in 1980, most of it for export, but in 1981 and 1982 this fell to just over 600 000 tonnes, rising again to nearly 800 000 tonnes in 1983.

Production is also expected to commence from a deposit discovered more recently near Port Loko, totalling around 30 million tonnes and averaging 49% Al_2O_3 and 4% SiO_2. In addition there are plans to mine bauxite near Kakata in Liberia.

3.9.4 Rutile

Placer deposits of rutile (TiO_2) are also of Tertiary to Quaternary age, the result of alluvial and eluvial processes that filled Tertiary valley systems. They are considered here because they originate from the charnockitic granulites and amphibolites of the Kasila Group.

The deposits consist in general of poorly sorted and weakly consolidated sands and clays, and rutile is the most important economic mineral in them. They occur all along the Sierra Leone coast, but the richest deposits are in the vicinity of Sherbro Island, where grades are of the order of 2–3% TiO_2. There is a major dredging and processing plant at Mogbweno, mining the coastal sands. Production began in 1979 and over 50000 tonnes of 99% TiO_2 concentrate were being shipped out annually up to 1983. Total production capacity is 100000 tonnes, which makes Sierra Leone a major world producer of this mineral, as well as having the world's largest proven reserves.

Several thousand tons of rutile-bearing alluvial deposits, derived from rocks of the Man complex in western Ivory Coast, locally achieve grades of 5–10 kg TiO_2 per cubic metre, and would be suitable for small-scale mining ventures.

It is worth mentioning here also the deposits of rutile-bearing sands on the coast near Banjul in Gambia, and further north in Senegal, where some 20 million tonnes are thought to be available. The origin of these deposits is not known because the coastal hinterland here is underlain by sediments of the Senegal Basin.

3.9.5 Other minerals

Chromite mineralisation is confined to the lower ultramafic units of the greenstone belts in the western part of the Archaean domain, in Sierra Leone. It occurs in alpine-type bodies, as impersistent bands and lenses from a few millimetres to a few metres thick, in sheared mafic and ultramafic (dunite–serpentinite) rocks. Average grade is of the order of 42–45% Cr_2O_3, which is too low for metallurgical-grade chromite (minimum 48% Cr_2O_3). The most important group of deposits is at Hangha in the northern Kambui Hills (Fig. 3.5), where a total of over 300000 tonnes was mined by open-pit mining up to 1963, when the low grade, small reserves and high transport costs made further extraction unprofitable. Chromite also occurs in the Gori Hills, but chromite occurrences in these belts are not sufficiently large or high grade to constitute ore deposits, unless mined by low-capital methods.

Nickel, asbestos and *talc* are also associated with ultramafic rocks, but no worthwhile deposits have so far been found. Lateritisation of ultramafic rocks can lead to secondary nickel enrichment and thick laterites in the northern Sula Mountains might have some potential. The asbestos minerals chrysotile and anthophyllite are widespread among the serpentinites of the greenstone belts, sometimes accompanied by talc, and some occurrences may prove to be of mineable quality. *Titanium*-rich magnetite occurs in mafic rocks of the Man complex in western Ivory Coast.

Traces of *molybdenite, sphalerite* and *galena* and signs of *copper* and *cobalt* mineralisation have been found in and around the greenstone belts, molybdenite and copper minerals in particular being associated with nearby granites, some of which could be palaeoporphyries. Small alluvial concentrations of *columbite* and *cassiterite* have been found in Sierra Leone, also derived from the granites, and there are placer deposits of *monazite* and other *rare-earth* minerals in western Ivory Coast. It is unlikely that important occurrences of any of these minerals will be found among the Archaean rocks of the West African craton. There may be some potential for *uranium* occurrences in the vicinity of granites, especially where they intersect the supracrustal belts, and uranium exploration has begun in both Liberia and Guinea. Economic minerals not normally expected among basement rocks include *barite*, mined on a limited scale near Kakata, some 50 km inland from Monrovia, and an unusual (probably residual) Fe–Al *phosphate* at Bambuta, containing over 1.5 million tonnes, from which it is planned to supply Liberia's fertiliser requirements.

Alumina-rich rocks of the Kasila belt may also constitute important concentrations of *kyanite* and *sillimanite* (in addition to bauxite; Sec. 3.9.3). In Liberia, kyanite–gneisses occupy an approximately 15 × 2 km strip of country near the town of Buchanan, with rocks containing up to 40% of kyanite.

Eventually it may also be possible to market the minerals associated with rutile in the coastal sands, mainly *ilmenite* and *zircon*, along with some *monazite* and *garnet*; though these are less plentiful and there are major sources in other parts of the world.

There is also some potential for extraction of relatively low-value resources from Archaean rocks. Most obvious is crushed rock for aggregate, from resistant granites and migmatites, but pure grades of quartzite and pegmatite–feldspar could be found and exploited for glass and ceramic manufacture.

Investigation into the mineral potential of the West African Archaean have not been so detailed as in other parts of the world. With the advent of sophisticated geophysical and remote-sensing techniques for exploration, it is likely that other economically viable deposits remain to be discovered.

Bibliography

Angoran, Y. and E. Kadio 1983. Aperçu de precambrien de Côte d'Ivoire: géologie–metallogenie. *J. Afr. Earth Sci.* **1**, 167–76.

Bárdosey, G. 1979. The growing significance of bauxites. *Episodes* **1979**, no. 2 (July).

Beckinsale, R. D., N. H. Gale, R. J. Pankhurst, A. McFarlane, M. J. Crow, J. W. Arthurs and A. F. Wilkinson 1980. Discordant Rb–Sr and Pb–Pb whole rock isochron ages for the Archaean basement of Sierra Leone. *Precambrian Res.* **13**, 63–76.

Behrendt, J. C. and C. S. Wotorson 1970. Aeromagnetic and gravity investigations of the coastal area and continental shelf of Liberia, West Africa, and their relation to continental drift. *Geol. Soc. Am. Bull.* **81**, 3563–74.

Behrendt, J. C., J. Schlee, J. M. Robb and M. K. Silverstein 1974. Structure of the continental margin of Liberia, West Africa. *Geol. Soc. Am. Bull.* **85**, 1143–58.

Béssoles, B. 1977. *Géologie de l'Afrique: le craton Ouest Africain.* Mem. BRGM Fr., no. 88.

Blanchot, A., J. P. Dumas and A. Papon 1973. *Carte géologique de la partie méridionale de l'Afrique de l'Ouest,* 1:2 000 000. Paris: BRGM.

Cahen, L. and N. J. Snelling 1984. *Geochronology and evolution of Africa.* Oxford: Oxford University Press.

Coney, P. J., D. L. Jones and J. W. H. Moyer 1980. Cordilleran suspect terranes. *Nature* **288**, 329–33.

Hawkes, D. D. 1972. The geology of Sierra Leone. In *African geology,* T. F. J. Dessauvagie and A. J. Whiteman (eds). Ibadan: Ibadan University Press.

Hurley, P. M., G. W. Leo, R. W. White and H. W. Fairbairn 1971. Liberian age province (about 2700 m.y.) and adjacent provinces in Liberia and Sierra Leone. *Geol. Soc. Am. Bull.* **82**, 3483–90.

Macfarlane, A., M. J. Crow, J. W. Arthurs, A. F. Wilkinson and J. W. Aucott 1981. *The geology and mineral resources of northern Sierra Leone.* Overseas Memoir 7, Institute of Geological Sciences, UK. London: HMSO.

Morel, S. W. 1979. The geology and mineral resources of Sierra Leone. *Econ. Geol.* **74**, 1563–76.

Nyema Jones, A. E. and W. E. Stewart 1972. General geology of Liberia. In *African geology,* T. F. J. Dessauvagie and A. J. Whiteman (eds). Ibadan: Ibadan University Press.

Olade, M. A. 1980. Precambrian metallogeny in West Africa. *Geol. Rdsch.* **69**, 411–28.

Papon, A., M. Roques and M. Vachette 1968. Age de 2700 million d'années, determine par la method au strontium, pour la série charnockitique de Man, en Côte d'Ivoire. *C.R. Acad. Sci. Paris, D* **266**, 2046–8.

Prasad, G. 1983. A review of the early Tertiary bauxite event in South America, Africa and India. *J. Afr. Earth Sci.* **1**, 305–14.

Rollinson, H. R. 1978. Zonation of supracrustal relics in the Archaean of Sierra Leone, Liberia, Guinea and Ivory Coast. *Nature* **272**, 440–2.

Rollinson, H. R. 1982. *P–T* conditions in coeval greenstone belts and granulites from the Archaean of Sierra Leone. *Earth Planet Sci. Lett.* **59**, 177–91.

Rollinson, H. R. 1983. The geochemistry of mafic and ultramafic rocks from the Archaean greenstone belts of Sierra Leone. *Mineral Mag.* **47**, 267–80.

Rollinson, H. R. and R. A. Cliff 1982. New Rb–Sr age determinations on the Archaean basement of eastern Sierra Leone. *Precambrian Res.* **17**, 63–72.

Vachette, M. and I. Yace 1969. Sur l'age au strontium des laves acides du Precambrien de la région de Roumodi en Côte d'Ivoire. *C. R. Acad. Sci. Paris, D* **268**, 2235–6.

Williams, H. R. 1978. The Archaean geology of Sierra Leone. *Precambrian Res.* **6**, 251–68.

Williams, H. R. 1979. An Archaean suture in Sierra Leone? *Nature* **282**, 608–9.

Notes: Up-to-date information on progress and production in various sectors of the mineral industry in West Africa is provided by the *Mining Annual Review*, published by Mining Journal Ltd, London. An extremely useful compendium of background information on mineral resources, relevant to virtually all chapters of this book, is provided in *Mineral Resources of Africa*, by Nicholas de Kun, published by Elsevier (1965).

4 *The Proterozoic of West Africa*

SUMMARY

This chapter covers the eastern portion of the West African craton, dominated by Birimian supracrustals that were deformed and metamorphosed in the Eburnian thermotectonic event about 2 billion years ago. The supracrustals were probably deposited at least partly upon an older Liberian basement, but evidence for this is not conclusive.

Both metavolcanic greenstone-type lithologies and metasediments occur among the Birimian sequences, the former predominating in the eastern half of the region, the latter in the west. In addition, there is another much less widespread assemblage of supracrustals, the Tarkwaian, which is important in some places for its gold-bearing potential.

Two contrasted views of Birimian–Tarkwaian stratigraphy have been put forward, probably because there are regional differences in the distribution of volcanic and sedimentary supracrustals. In Ghana, the Lower Birimian is identified as consisting of metasediments, the Upper Biri- mian of metavolcanics, and the Tarkwaian rocks are considered to lie unconformably on the Birimian. In Ivory Coast and Upper Volta, no such clearcut subdivisions are recognised. The metavolcanics are considered to be either older than or broadly contemporaneous with the metasediments, and the Tarkwaian rocks are regarded as lateral facies variants of the Birimian sediments.

Many large concordant syntectonic batholithic granite masses occupy the antiformal areas between Birimian synforms, and smaller late-tectonic granites are also numerous. Some of the larger bodies may be remobilised Liberian or older basement.

The main economic mineral deposits in this part of the craton are gold, manganese, diamonds and bauxite. There are several mineral deposits of lesser importance, including iron ores and pegmatite deposits of tin and related elements.

4.1 Introduction

Lower Proterozoic rocks representing the timespan from about 2500 to 1800 Ma form the major part of the West African craton (Guinea Rise). The Birimian sediments and volcanics accumulated after the Liberian event, over an area that covered what is now Ghana and Ivory Coast, much of Upper Volta, and parts of northern Guinea, southwestern Mali, southeastern Senegal, western Niger and southeastern Liberia. They may extend north (beneath the younger sedimentary cover) as far as Morocco (Fig. 1.3), and they may also be represented on the other side of the Atlantic in north-east Brazil and Guyana, which are considered to have been continuous with West Africa when this region was part of Gondwanaland.

Figure 4.1 shows the proportion of supracrustals relative to basement to be a great deal higher in the Lower Proterozoic terrane than in the Archaean (Fig. 3.1). The area of crust affected by the *c*. 2000 Ma Eburnian thermotectonic event is about 1000 km wide, from eastern Ghana to eastern Liberia. The regional trend of the rocks within the area of Figure 4.1 is similar to that of the Liberian (Fig. 3.1), i.e. between N–S and NE–SW. This is at least consistent with the existence of an older basement, the structural grain of which controlled subsequent tectonic trends. Folding is mostly tight to isoclinal, with a steeply dipping foliation (70–90°) that is generally parallel to lithological layering.

Fractures and faults are both parallel and perpendicular to the strike of foliation, a pattern similar to that in the Archaean terrane (Sec. 3.5). There are also some major N–S fractures and some of these contain dolerite dykes, which may, however, be younger than Eburnian.

The whole region underlain by Lower Proterozoic Birimian supracrustals and older basement affected by regional deformation, metamorphism and granite emplacement in the Eburnian event is also known as the **Baoulé–Mossi domain**. Although lying outside the main area of the domain, the Kayes and Kenieba inliers in the north-west (eastern Senegal) may be considered to belong to it. The eastern boundary of the Baoulé–Mossi domain with the younger Pan African domain to the east is marked by the thrust zones of the Togo belt (Ch. 6). In the west, the boundary with the older Kenema–Man domain is less well defined, except along the Sassandra mylonite zone.

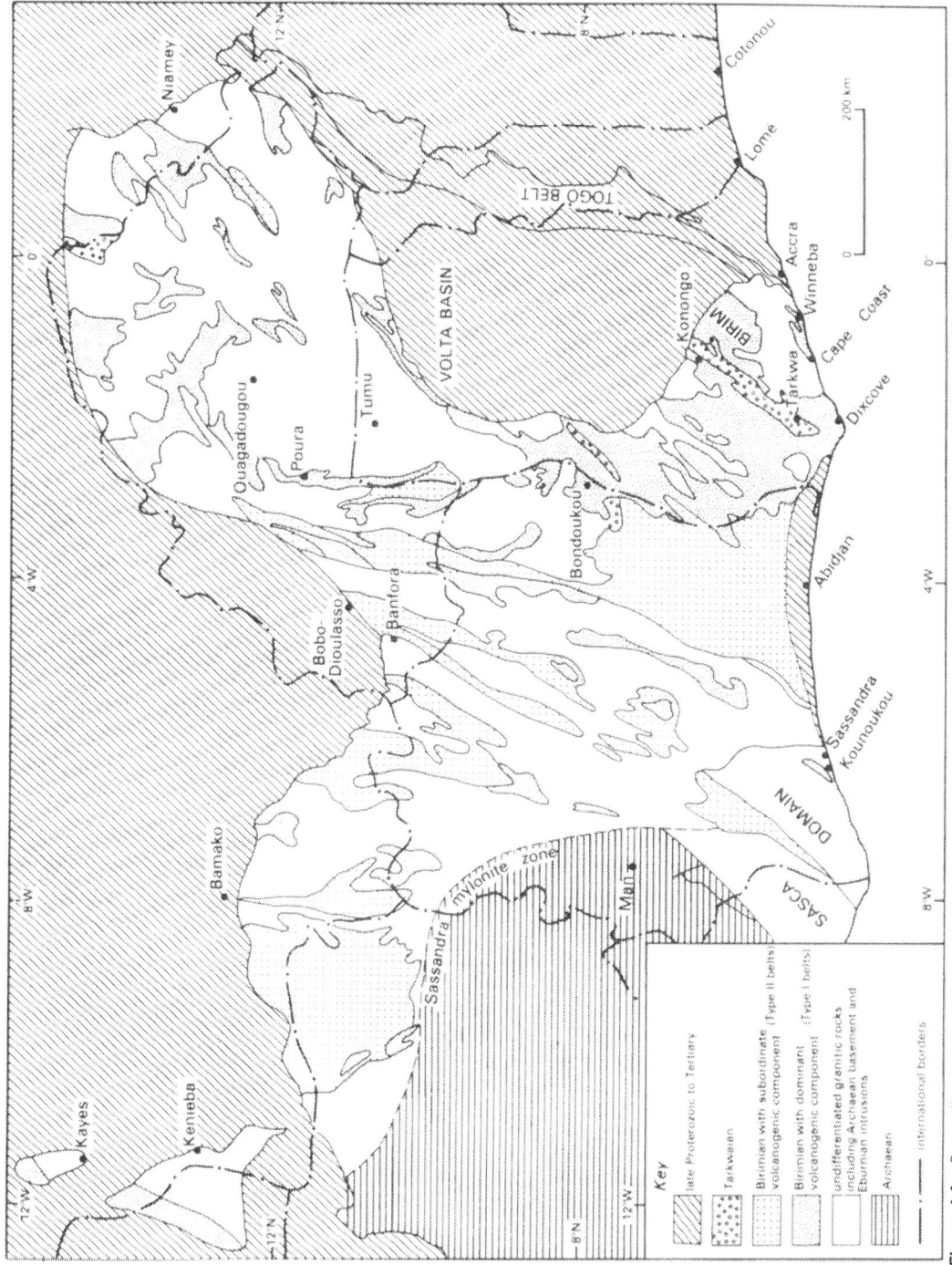

Figure 4.1 Generalised distribution of Birimian supracrustal belts in West Africa.

Key

Late Proterozoic to Tertiary

Tarkwaian

Birimian with subordinate volcanogenic component (Type II belts)

Birimian with dominant volcanogenic component (Type I belts)

undifferentiated granitic rocks including Archaean basement and Eburnian intrusions

Archaean

international borders

200 km

VOLTA BASIN

TOGO BELT

BIRIM

SASCA DOMAIN

Sassandra mylonite zone

Niamey

Ouagadougou

Poura

Tumu

Bobo-Dioulasso

Banfora

Bamako

Keneiba

Kayes

Man

Kounoukou

Sassandra

Abidjan

Bondoukou

Konongo

Tarkwa

Cape Coast

Winneba

Accra

Dixcove

Lome

Cotonou

4.2 Basement

There is no unequivocal evidence of older basement in the Proterozoic Baoulé–Mossi domain, but it is very likely that much of the undifferentiated gneiss, migmatite and granite terrane is reactivated Archaean crust, occupying the antiformal areas ('*moles*' of French geologists) between the synformal supracrustal belts. Several basement areas have been identified as retaining some evidence of pre-Birimian events throughout the Baoulé–Mossi domain. These are not distinguished in Figure 4.1, except for the largest which is in southwestern Ivory Coast, and it may be that this region should properly be included with the Archaean nucleus. It is sometimes referred to as the **Sasca domain** (cf. Fig. 2.1), and its boundary with the undisturbed Archaean of the Kenema–Man domain (Fig. 3.3) coincides with a marked north-east trending break in the geomagnetic fabric of coastal Liberia.

As in the Archaean nucleus, no convincing sedimentary contacts between basement and supracrustals have been described, but clasts of gneiss, migmatite and granite occur in Birimian sediments in places, for example near Banfora in Upper Volta and in southwestern Niger (Liptako region), and thus support the contention that older continental basement was present. Structural and metamorphic discordances between lower-grade Birimian and the higher-grade gneissic rocks are common, and many boundaries are faulted or sheared. Migmatites and gneisses in eastern Upper Volta, southwestern Niger and northern Ghana approach granulite facies metamorphic grades, but this is not necessarily evidence of greater age. A hill of ferruginous and manganiferous quartzite near Winneba, west of Accra, may represent an Archaean supracrustal relic, implying the existence of older basement in at least part of southeastern Ghana. Migmatites in northeastern Ivory Coast and granitic gneisses near Bobo-Dioulasso and Poura in Upper Volta could also be pre-Birimian.

The largest and best-studied area of probable pre-Birimian basement is the Sasca domain in southwestern Ivory Coast, extending across the border into southeastern Liberia. Here are supracrustal ferruginous and manganiferous quartzites, with amphibolites and pyroxenites, among granitic and pelitic gneisses and migmatites ranging up to granulite facies. Three phases of deformation are identified in the small belt of Birimian schists and paragneisses near Kounoukou. These are also found in the surrounding rocks, superimposed on an earlier deformation, probably Liberian.

4.3 Supracrustals: the Birimian and Tarkwaian

The very thick and extensive sequence of metamorphosed sediments and volcanics that dominates this age province is called the **Birimian** after the Birim region in southern Ghana where the rocks were first described in detail. Metamorphic grades range from greenschist to almandine–amphibolite facies in these rocks, which are an important source of diamonds and manganese ores (Sec. 4.7).

A much smaller and more scattered group of supracrustals, mainly shallow-water sediments, is called the Tarkwaian (after the town of Tarkwa in southern Ghana, where they are gold-bearing). Although greenschist to almandine–amphibolite facies metamorphism is recorded from these rocks, in many places they are described as being 'hardly metamorphosed'. A consensus has not yet been reached about the stratigraphy of the Birimian and its relationship to the Tarkwaian. Geologists working in Ghana on the one hand, and Ivory Coast and Upper Volta on the other, have arrived at different conclusions. It will be necessary to deal with the two views separately and then to examine reasons for the differences between them.

4.3.1 *The Birimian in Ghana*
In Ghana there is a long-established subdivision into Lower Birimian, dominated by metasediments, and Upper Birimian, dominated by greenstone-type metavolcanics. These are not distinguished on Figure 4.1.

The lowest parts of the succession are primarily phyllites and greywackes. These change upwards to phyllites and weakly metamorphosed tuffs, greywackes and feldspathic sandstones, and the sequence appears to pass conformably into the Upper Birimian, although a local unconformity has been recorded in places. Some of the phyllites contain pyrite, and finely divided carbonaceous matter is present in most of them. Silicification is common among the phyllites, particularly towards the boundary with the Upper Birimian. Quartzites, calcareous rocks and conglomerates are rare, but the conglomeratic horizons contain fragments of granitic and other rocks believed to be derived from older basement.

The Upper Birimian consists chiefly of meta-morphosed basaltic and andesitic lavas, now hornblende–actinolite–schists, calcareous chlorite-schists and amphibolites (the greenstones). Pillow structures indicating sub-aqueous eruption of the original basaltic lavas are frequently observed. Minor intrusions of mafic rocks cut the volcanics and there are small ultramafic bodies in some places. Smaller amounts of rhyolitic and dacitic lavas and tuffs are also recorded, and subordinate metasedi-ments include phyllite, greywacke, quartz–sericite-schists and mica–schists, as well as grits and con-glomerates at the base of the succession.

Bands of **gondite** (quartz–spessartite rock) and manganiferous phyllite occur within the greenstones towards the top of the Upper Birimian, commonly associated with tuffs, silicified argillites (hornstones) and chert. The rocks are dark and finely banded, and the presence of carbonaceous matter in them suggests deposition in still waters deficient in oxy-gen. Mn-rich horizons also occur at stratigraphically lower levels in the Upper Birimian and have been found in uppermost Lower Birimian as well.

Because the rocks are tightly folded and com-monly sheared and fractured, it is not easy to establish stratigraphic successions and estimate thicknesses. The total thickness of the Birimian in Ghana may be of the order of 10 000 to 15 000 m.

4.3.2 The Tarkwaian in Ghana

The main occurrences of **Tarkwaian** sediments occupy two generally synclinal belts surrounded by Upper Birimian metavolcanics, about 270 km apart, and smaller occurrences of Tarkwaian rocks may occur elsewhere. The sediments are mainly of shallow-water origin, probably fluviatile, and they contain fragments of Birimian rocks. They could have been deposited in separate elongate basins as molassic facies derived from erosion of the Birimian, during later stages of the Eburnian orogeny.

In Ghana, there is in places a strong angular unconformity between the Birimian and Tarkwaian, and the Tarkwaian appears in general to be less strongly deformed and metamorphosed than the Birimian. However, neither the unconformity nor the difference in intensity of folding and metamorphism can be convincingly demonstrated at all outcrops. The overall synclinal structure of the Tarkwaian in the vicinity of Tarkwa itself is rela-tively simple, with fairly open folds having a north-easterly plunge and northwesterly dipping foliation, though the intensity of folding increases to the north-west. In the Bui syncline, Tarkwaian beds have been considerably fractured and more strongly folded, being overturned in places.

According to the Ghanaian 'school', therefore, Upper Birimian metavolcanics occupy the cores of the rather broad Birimian synforms, with granitic rocks (partly basement) in the intervening antiforms (Fig. 4.1). Tarkwaian rocks form the central parts of two synformal belts, and in the larger Tarkwa syncline an unconformity is implied by the fact that the Tarkwaian lies mainly against Upper Birimian metavolcanics in the west, but overlap onto Lower Birimian in the east.

Although the Birimian and Tarkwaian of Ghana may have been studied for longer and in more detail than in adjacent countries, they form only a small part of the Proterozoic domain and are not typical of the whole of it.

4.3.3 The Birimian and Tarkwaian in Ivory Coast and Upper Volta

In northeastern Upper Volta (and southwestern Niger), the supracrustal belts consist mainly of greenstone sequences and volcanosedimentary suc-cessions similar to those in Ghana, though they are in general smaller (Fig. 4.1). These are sometimes called **Type I belts** by francophone geologists, to distinguish them from the **Type II belts** which occupy most of the western part of the Proterozoic terrane, that is, Ivory Coast (and northern Guinea). These are mainly sediments of shallow-water origin, among which quartzites, mica–schists, metagrey-wackes and metaconglomerates are important litho-logies. Calc-silicate rocks also occur, and volcanic rocks are subordinate, ranging from mafic through intermediate to acid compositions.

In Ivory Coast and Upper Volta, the greenstone facies (Upper Birimian of Ghana) is generally con-sidered to be either older than or broadly con-temporaneous with the predominantly sedimentary facies (Lower Birimian of Ghana). These interpreta-ions are based partly on the occurrence of volcanic pebbles in conglomerates of the sedimentary forma-tions. Mica–schist pebbles are also found in some intraformational conglomerates, implying that there was deformation, metamorphism and erosion between the deposition of Lower and Upper Biri-mian sequences in at least some places (Sec. 4.5).

The 'Tarkwaian' also occurs in Ivory Coast and Upper Volta, as scattered sequences of molasse-type sediments, with or without conglomerates. The two main occurrences, in eastern Ivory Coast and north-east Upper Volta, are shown on Fig. 4.1. Their relationship to the Birimian is less simply defined than in Ghana, because the two have been interfolded to produce very complex structures. In consequence, sediments of Tarkwaian type seem to be regarded as lateral facies variations within the main Birimian sedimentary facies: a series of local shallow deposits of roughly similar character, occurring at various stratigraphic levels, but characterised by unconformable relationships with underlying rocks. Near Bondoukou, for example, the rocks are stated to overlie both Birimian and granite; dips are low but the rocks have suffered heavy crushing, possibly as a result of thrust movements.

4.3.4 A possible compromise

Figure 4.2 summarises the principal features of the Ghanaian and Ivory Coast/Upper Volta 'schools' concerning Birimian–Tarkwaian relationships. In Ghana, the volcanic rocks are interpreted as representing very widespread igneous activity following the accumulation of a thick sedimentary sequence.

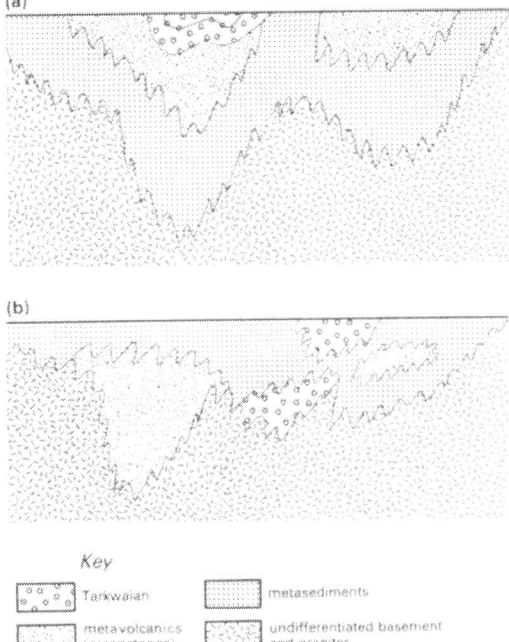

Figure 4.2 Diagrammatic cross sections illustrating Birimian and Tarkwaian relationships as envisaged in (a) Ghana and (b) Ivory Coast and Upper Volta.

In Ivory Coast, however, the volcanics are seen as having been erupted along zones of weakness near the borders of depositional basins, so that some greenstones are older than the metasediments, others more or less of the same age, and the Tarkwaian rocks are simply lateral facies variants within Birimian metasediments.

The radically different interpretations summarised in Figure 4.2 could be ascribed to a combination of logistic and geological factors, which apply to other regions too. Outcrops and communications are generally poor throughout much of the region, and geological mapping is still mainly at reconnaissance level. Foliation of the supracrustals is almost everywhere steep and parallel to original sedimentary layering, and sedimentary structures that would give way-up indications are not always easy to find. Structures are often complex and, where different groups of rocks have been folded together, their relative ages and original relationships are difficult to unravel – unconformities can be obliterated for example, as they have been in this domain. Under these circumstances, geologists working in different regions could well reach conflicting conclusions about the stratigraphy and correlation of rock units.

However, it seems likely that there are more fundamental geological reasons for the different interpretations. Figure 4.1 shows that there are far more Birimian volcanics in the eastern half of the Baoulé–Mossi domain than in the west, and there is no a priori reason why they should all be contemporaneous over the whole of this vast region.

Volcanism was obviously more sporadic and scattered in the west (Ivory Coast, northern Guinea), and could have broken out at any time. The more voluminous volcanic activity in the east could have begun earlier in the north (eastern Upper Volta, southern Niger) than in the south (Ghana). An explanation along these lines offers a plausible way of reconciling the contrasted Birimian stratigraphies that have been established in different parts of the Proterozoic domain.

4.4 Granitic rocks

Most of the granites in the region of Figure 4.1 fall into two main groups:

(a) Large syntectonic batholithic granites have been designated the **Cape Coast type** in Ghana and the **Baoulé type** in Ivory Coast and Upper

Volta. They are generally concordant with regional structures and are often foliated. Many are two-mica granites, though biotite- and hornblende-bearing varieties are also common. Granodioritic compositions predominate, along with K-rich microcline–phyric adamellites. These larger concordant granitic masses may have metamorphic aureoles where they intrude the supracrustals, but elsewhere they are often migmatitic round their margins and exhibit structural characteristics of the surrounding rocks. They were probably derived in part at least by remobilisation of older Liberian basement, perhaps in part also by granitisation and partial melting of Birimian metasediments. Pegmatite facies occur throughout these concordant batholiths, sometimes reaching several metres across, and microgranite and aplite veins are also common.

(b) Smaller discordant and typically unfoliated late-tectonic to post-tectonic granites, often with subcircular outcrops, are designated the **Dixcove type** in Ghana and the **Bondoukou type** in Ivory Coast and Upper Volta. They are less abundant than the older syntectonic granites and have a wider compositional range: from hornblende- and biotite-bearing granites to diorites, monzonites and syenites.

In Ghana, there is some controversy as to whether any of the Eburnian granitic intrusions are post-Tarkwaian. Granites adjacent to Tarkwaian sediments (as near Konongo in Ghana) appear to have structurally and metamorphically influenced Birimian rocks but not the Tarkwaian. Aplitic veins related to these granites are found cutting Birimian but not Tarkwaian rocks, and granite pebbles occur in Tarkwaian sediments. On the other hand, pegmatites have been found cutting the Tarkwaian in other places (Sec. 4.5).

However, there are basic to intermediate intrusive rocks among the Tarkwaian sediments, so there was clearly some igneous activity in progress when they were deposited. Igneous rocks of similar composition and age probably occur elsewhere. The largest basic intrusions in Ghana, for example a norite body about 20 km long near Tumu in the north of the country, could also be a manifestation of late Eburnian magmatism. Some of the dykes referred to earlier (Sec. 4.1) also belong to this phase, for in places they are older than post-tectonic granites and have been affected by Eburnian metamorphism.

4.5 Correlation and geochronology

Because of the different stratigraphic interpretations outlined in Section 4.3, correlation throughout the Baoulé–Mossi domain can only be rather general. Over most of the area the Tarkwaian is generally regarded merely as a local variant of the Birimian, which can accordingly be simply subdivided into Upper and Lower. An initial phase of deformation and metamorphism along with emplacement of Baoulé-type syntectonic foliated granites has been identified as affecting only Lower Birimian rocks. This is called the Eburnian I phase. Pebbles of mica–schist and igneous rocks in Upper Birimian conglomerates (Sec. 4.3.3) were derived from the erosion of rocks formed in this deformation. The Eburnian II deformation and metamorphism was accompanied by early emplacement of some more syntectonic Baoulé-type granites and a later episode of post-tectonic intrusions, including the Bondoukou-type granites (Sec. 4.4).

In Ghana, with a three-fold subdivision (Upper and Lower Birimian, and Tarkwaian), this scheme is less easily accommodated. The Eburnian I could be correlated with the unconformities identified locally between Lower and Upper Birimian (Sec. 4.3.1) and Eburnian II would then pre-date the deposition of Tarkwaian sediments. These can in any case be regarded as a molasse facies, eroded from the deformed Birimian supracrustals, and themselves deformed in the waning phases of the Eburnian event. The evidence that most granite emplacement was over by the time the Tarkwaian sediments in Ghana were deposited (Sec. 4.4) is certainly consistent with this interpretation.

Whether or not the subdivision into Eburnian I and II events can be justified on the available field evidence, it is not yet possible to use geochronological data to define it. Numerous age determinations have been made on a variety of minerals and rocks, including metasediments (mica–schists), metavolcanics (acid lavas), granites and migmatites, by K/Ar, Rb/Sr and U/Pb methods. They provide a consistent pattern of ages for the Eburnian thermotectonic event throughout the area of Figure 4.1, falling within or close to the range of 2150–1950 Ma. It is likely that the climax of deformation, metamorphism and granite emplacement occurred at around 2100 Ma, i.e. in the later part of Lower Proterozoic time. In terms of the Eburnian I–II division, this would be the age of Eburnian II, which presumably overprinted most of the structures and other evidence of the earlier Eburnian I event, estimated to have occurred at around 2300 Ma.

In Ghana, folding and metamorphism of the Tarkwaian probably occurred after this climax. The sediments contain granite pebbles, but are not themselves intruded by granites (Sec. 4.4). However, a pegmatite cutting Tarkwaian rocks has given an age of 1650 Ma, which is taken as dating the final phase of deformation and igneous activity in this region (it must post-date the basic sills in the Tarkwaian as these were folded and metamorphosed with the sediments).

As already noted, there are no undisputed relict Liberian ages recording the presence of a pre-existing basement in this region, except in the transitional zone to the cratonic nucleus, the Sasca domain. Evidence for such a basement remains circumstantial, therefore, and it must be concluded that pre-Birimian rocks were substantially homogenised and had their radioactive 'clocks' reset by the Eburnian event.

4.6 Plate tectonics and the Eburnian event

If the transition from Archaean to Proterozoic represents a major change in the patterns of global tectonics (Sec. 1.2), this change is not immediately obvious among the Precambrian terranes of West Africa, where the basement–supracrustal–granite association is remarkably similar throughout, from Archaean to Pan African domains (Fig. 2.4).

So far as the Lower Proterozoic Birimian region is concerned, the only significant difference from the Archaean lies in the area underlain by supracrustals, which is much greater in the Birimian – though even this could be simply a function of the level of erosion. Otherwise the differences are not great. There are some lithological contrasts, such as the relative proportions of iron-rich and manganese-rich rocks among the metasediments. These apart, the broad distribution of Birimian lithologies approximates to a mirror image of the Archaean supracrustal belts. Greenstone-dominated (Type I) belts characterise the eastern half of the Baoulé–Mossi domain, metasediment-dominated (Type II) belts the western half. The regional north to north-east trend is characteristic of most Birimian belts, as it is in the Archaean, and the swing to northwesterly trends in the north of the Archaean nucleus (Fig. 3.1) is also seen in the adjacent Birimian belts of northern Guinea (Fig. 4.1).

Plate tectonic models have been proposed for Lower Proterozoic terranes in other parts of the world, such as the Canadian shield, where basement–supracrustal–granite associations can also be identified. The Lower Proterozoic of West Africa is insufficiently known for similar models to be proposed with any confidence, but some general observations are possible.

The Sassandra mylonite zone may approximate to a crustal suture along the eastern margin of the cratonic nucleus. At least some of the granitic areas in Figure 4.1 presumably represent areas of older continental crust, within and between which were ensialic and ensimatic volcanosedimentary basins, now represented by the supracrustal belts. The abundance and diversity of Birimian supracrustal belts and the great size of the Eburnian domain suggests that the Eburnian event was the culmination of many minor collisions involving the aggregation of continental blocks and the closure of numerous back-arc and inter-arc basins. These may have been predominantly ensialic in the west of Figure 4.1, predominantly ensimatic in the east, judging from the distribution of Type I and II supracrustal belts.

If the Birimian continental crust formed by mechanisms of the kind summarised in Figure 2.6, one important constraint is the direction of subduction zones marginal to the Archaean nucleus. These must have been directed away from the margin, or the Liberian domain would probably have been more extensively remobilised in the Eburnian.

Alternatively, the Birimian supracrustals were mainly deposited in ensialic basins developed over a large area of continental crust and deformed mainly by differential vertical movements that involved relatively little crustal shortening. Indeed, it is entirely possible that some Birimian belts originated in this way, and geophysical data have been used as a basis for models dominated by rifting. Such a mechanism does not preclude the operation of plate tectonics, for the rifted basins could have formed above shallowly inclined subduction zones, as in the Andean belt of western South America.

Whatever the tectonic mechanism proves ultimately to be, however, it should be remembered that the present limits of the Archaean and Proterozoic age provinces do not represent the extent of ancient crust in West Africa. There is good evidence that both the Liberian and Eburnian events influenced the crust further east, in the Pan African age provinces (Fig. 2.1). Together with the existence of numerous presumed pre-Birimian relics in the Baoulé–Mossi domain, this supports the contention that much of the West African continental crust was in existence at the end of the Liberian (Sec. 3.8), even though the different segments were not necessarily in their present-day position. Another powerful argument for the existence of ancient continental crust in this region is the occurrence of diamonds in Birimian metasediments (Sec. 4.7.3). The diamonds can only have come from kimberlites emplaced in stable and relatively cool continental crust, and such crust must have been present here about 2500 Ma ago.

4.7 Economic potential in the Lower Proterozoic rocks of West Africa

This region is a more promising metallogenic province than the Archaean cratonic nucleus, principally because of the vast area of supracrustal rocks preserved.

Figure 4.3 shows the distribution of some principal mineral deposits in the Lower Proterozoic part of the craton. Those of major importance are gold, manganese, diamonds and bauxite. Most of the mineralisation is in Type I Birimian greenstones and the Tarkwaian rocks, or in soils and gravels above these formations. It is strongly tectonically controlled, most deposits lying within or perpendicular to the regional structural grain. There are many more smaller occurrences than those shown in Figure 4.3, which is intended to provide a general idea of the mineral potential.

4.7.1 *Gold*

The major auriferous zones of West Africa are located in Ghana, known as the Gold Coast till independence in 1957. It has been estimated that the Gold Coast provided two-thirds of Africa's gold production between the late fifteenth and mid-nineteenth centuries, and Ghana is still a major gold producer. Annual production is around 400 000 to 500 000 oz (*c.* 12 000–15 000 kg), though it was of the order of 10,000–11 000 kg in the early 1980s. Total reserves are estimated to exceed 5000 tonnes of exploitable gold, and annual production is planned to reach 50 000 kg by 1990, and to rise further thereafter.

Most of the primary gold deposits are located along the Lower–Upper Birimian boundary on the west side of the Tarkwa syncline, where three presently active mines are situated (Prestea, Obouasi and Konongo). Other occurrences in similar settings are on the east side of the Tarkwa syncline at its northern and southern ends, and further west in a belt extending from Sewum in the south to beyond Bibiani in the north.

Primary gold occurs elsewhere in Ghana too, and in Ivory Coast (e.g. Hire), Upper Volta (e.g. Poura) and in southwestern Mali, both in the main Birimian area and in the Kenieba inlier (Kalana), and there is also gold in eastern Senegal (Sabodala), where the Kenieba inlier extends into it.

In Ghana, the major primary gold lodes are associated with persistent and deep-seated shear zones that may have been partly controlled by local unconformities between Lower Birimian phyllites and Upper Birimian greenstones. The country rocks in general comprise metamorphosed carbonaceous and manganiferous argillites, tuffs and greywackes, along with basic to intermediate igneous rocks. The primary gold occurs in quartz veins and lenticular reefs and also in some of the tuffaceous and argillaceous rocks. It is accompanied by sulphides, especially arsenopyrote but including pyrite and pyrrhotite, chalcopyrite and bornite, and a little galena and sphalerite. There is up to 10% silver in the gold and this helps to pay for refining the ore.

Primary gold in the Birimian of other parts of West Africa occurs in generally similar settings to those in Ghana, and significant deposits are reported in Ivory Coast. The gold is probably of syngenetic volcano-exhalative origin, related to greenstone volcanism and associated sedimentation, remobilised during the Eburnian metamorphism to become concentrated and localised along major shears. Although there is no direct evidence of any spatial or temporal relation to granitic intrusions, emplacement of the granites may have provided an additional heat source for mobilisation of the gold into veins.

The vertical extent of the primary gold deposits is known to be several hundred metres, but difficulties of mining have meant that exploration below about 100 m has not been extensive. It is clear that this is worth while, as ore in the large Obuasi mine did not become especially rich until depths of 300 m were reached.

In Ghana, the greatest activity in gold exploration and mining occurred in the late nineteenth and early twentieth century but very few records were kept. It is virtually impossible to relocate earlier prospects, which might have been abandoned as uneconomic, when poor communications made several ore bodies

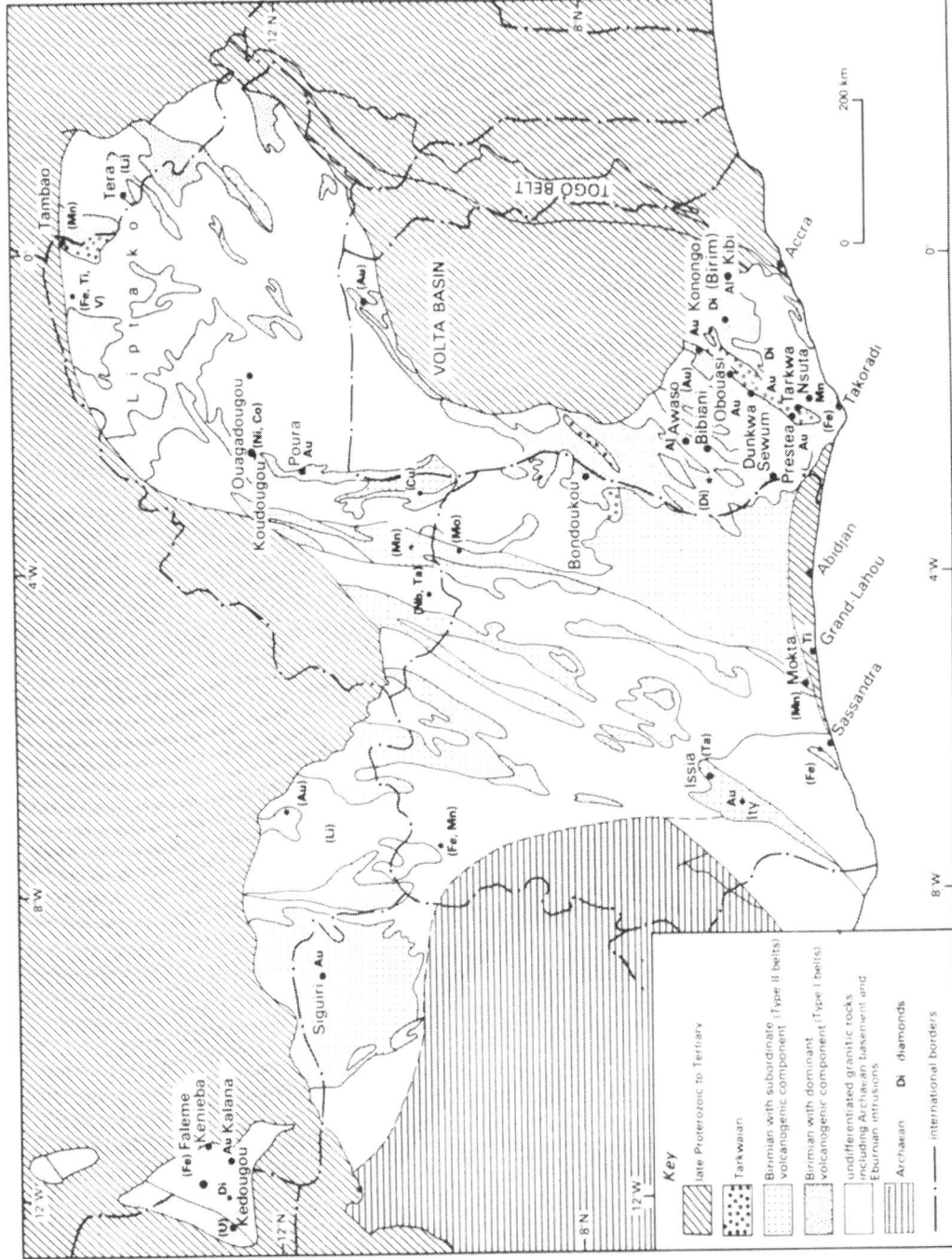

Figure 4.3 Distribution of some major economic mineral deposits associated with Lower Proterozoic rocks of the West African craton. Elements are shown by their chemical symbols. Symbols in brackets denote deposits that are known or suspected but not presently exploited.

of various sizes unprofitable. They were rich ores (20–30 ppm Au) and would be highly profitable today, with improved technology and the high price of gold.

The deposits need not be large in terms of total tonnage. For example, the reopened Kalana mine in Mali has reserves variously estimated at between 24 and 40 tonnes of gold in ore averaging 30 ppm Au. Initial production is 400 kg per year, rising to 2000 kg annually in the long term. At the Poura mine in Upper Volta, when production ceased in 1966 there were still some 1.5 million tonnes of ore left, averaging around 15 ppm Au and containing some 22 tonnes of gold in total. This mine is also being reactivated. With annual production of the order of 1 tonne of metal, such deposits have substantial lifetimes and exploration may in any case extend their reserves. Prospecting continues for gold in several parts of the Birimian domain.

Gold in the Tarkwaian Sedimentary gold is found in several places in the so-called Banket conglomerates near the base of the Tarkwaian, mainly along the eastern margin of the syncline, decreasing rapidly in abundance towards the west. It is associated with heavy minerals, including rutile, zircon and plentiful detrital haematite (often as much as 20%, and up to 60% in small pockets), which makes the Tarkwaian a major feature on aeromagnetic maps, though there is not enough to be economic. The heavy minerals and the gold are concentrated in the more mature and better sorted matrix-poor gravel horizons within the Banket. These gravel layers are not stratigraphically equivalent from place to place, but they are quite extensive, with ore grades remaining consistent over hundreds of metres. Diagenesis during folding of the Tarkwaian has resulted in recrystallisation of haematite and the introduction of some pyrite. The distribution of the gold and sedimentological studies of well preserved features such as graded and current bedding suggest that the sediments represent piedmont fan systems of braided and meandering stream channels, with an upland source lying mainly to the east.

Although gold might be expected to occur in similar quantities in the Bui syncline and other occurrences of Tarkwaian rocks (e.g. Bondoukou in Ivory Coast), this has not been found to be the case. Only traces of gold have been discovered, and there are no concentrations of haematitic black sands either.

Alluvial (placer) gold Modern stream channels near primary and secondary gold districts in the Birimian and Tarkwaian contain placer gold. The Ofin River system currently provides most of the placer gold extracted by dredging in Ghana, the main operation being at Dunkwa. In addition there are eluvial deposits, beach sands, terrace deposits and older Pleistocene stream sediments containing alluvial gold. Many such occurrences have been prospected and worked in the past and probably still are. There are notable alluvial concentrations in southeastern Liberia, northeastern Guinea (Siguiri) and southwestern Ivory Coast, on the Sassandra River system, where the Ity gold deposit is also situated.

4.7.2 Manganese

In Ghana manganese ores were discovered in 1914 at Nsuta in southern Ghana, south-east of Tarkwa. It was the richest manganese ore discovery in the world at that time. In addition to metallurgical-grade ore, the deposit contained large quantities of unique battery-grade ore called **nsutite**, which is almost pure MnO_2 and could be used in dry cells without processing. There are manganese ores elsewhere in Ghana, but the Nsuta deposit is by far the largest and is the only one exploited. Production in both 1980 and 1981 was over 220 000 tonnes, and was a little over 175 000 tonnes in 1983.

The ore occurs over a range of low hills in the Nsuta area and the manganese originates mainly from Upper Birimian manganiferous phyllites or their more highly metamorphosed equivalents, gondites (quartz–spessartite rocks). These primary rocks are uneconomic as sources of manganese. Deep weathering on a Tertiary peneplain resulted in supergene enrichment to produce almost pure manganese oxides. These constitute the rich ore and are approaching exhaustion. There appear to be large tonnages of carbonate ore, however, averaging 30–35% manganese, but evidently sufficiently enriched in places to give estimated reserves of some 10 million tonnes of about 48% Mn, which is still economically workable.

Manganiferous rocks outcrop frequently along both sides of the Tarkwa syncline and many of the deposits are in Upper Birimian rocks, though some in northwestern Ghana appear to be near the Lower–Upper Birimian boundary, which is also the locus of most of the primary gold mineralisation. The rocks comprise black slates as well as phyllites and gondites, along with basic and acid volcanics, including tuffs. The sediments are often markedly carbonaceous. The oxide ores have nodular or columnar and radiating forms near the surface, grading down into layered ores, where carbonate becomes more important. The main ore minerals are pyrolusite, nsutite, psilomelane and rhodochrosite, along with rhodonite and spessartite, and pyrite is common in the carbonate ores and intercalated argillaceous rocks.

The carbonate ores show signs of sedimentary layering, but it is not certain whether they are primary or derived by some form of chemical replacement from the manganiferous phyllites and gondites. The origin of the manganese remains undecided, though the frequent association with volcanic rocks suggests that it may have had a submarine volcano-exhalative origin, formed on an ancient sea bed much in the way that manganese-rich muds form on the floors of present-day oceans. Manganese and gold often occur together in the Ghanaian Birimian, which suggests the possibility that manganese could be used as a pathfinder element in exploration for gold.

Very similar manganese ores are found in southern Ivory Coast (Mokta and the Blafa-guéto hills), northeastern Upper Volta (Tambao) and in southwestern Niger. They were also derived by secondary enrichment from Birimian phyllites, gondites and carbonates (locally associated with rhyolites) during the Tertiary. They occur as residual cappings on low hills and ridges, as at Nsuta. The Tambao deposits have been estimated to contain some 13 million tonnes of 54% oxide ore and a similar amount of carbonate ore. Exploitation of this deposit will require the construction of a rail link to Ouagadougou some 360 km away, and concentrates would then have to be transported a further 1000 km or so, to the port at Abidjan. As a consequence, production may not commence till about 1990. There appear to be no plans at present to exploit other known manganese ores formed from Birimian rocks.

4.7.3 Diamonds

Alluvial diamonds have been found in Birimian rocks in Ghana and in Ivory Coast. Production comes mainly from industrial dredging, and to a minor extent from small-scale licensed digging of surface eluvial and alluvial placers in the overlying and adjacent soils and sediments. Ghana is the major source of diamonds produced from Birimian rocks, followed by Ivory Coast. Annual production of diamonds in Ghana was about 1–1.5 million carats during the 1970s, but by 1983 this had fallen to less than 340 000 carats. The diamonds are mostly small industrial stones, though with a significant proportion of gem quality. They come mainly from the Birim field of southeastern Ghana, the largest single diamond-producing area in West Africa. Here the immediate source of the diamonds is a band of Lower Birimian conglomerates (which may be coarse turbidite deposits) about 200 m wide and some 80 km long. Diamonds also occur further west in Ghana and have been reported from southeastern Ivory Coast as well, though these are generally smaller than the Birim diamonds. In the small Bonsa field, about 20 km south-west of Tarkwa, diamonds occur in conglomerates at the base of the Tarkwaian.

As the diamonds occur within Birimian and Tarkwaian rocks that are some 2000 Ma old, the source must be even older. No kimberlites have so far been found in Ghana. They may have been destroyed during the Eburnian deformation or metamorphism or they may lie concealed beneath the Voltaian sediments to the north; alternatively they are simply not exposed and so have not been located. The decrease in size of diamonds in the Birimian away from the major Birim field suggests that the source lay in or near this region and the diamonds were transported from it to the other fields, although this would require unusually large-scale transport for a mineral as dense as diamond.

Kimberlites occur in other parts of West Africa, notably in Sierra Leone and Liberia, which are important diamond producers, but also in northern parts of Ivory Coast and Ghana, in Guinea and in southwestern Mali. However, these kimberlites cannot have been the source of alluvial diamonds within the Birimian as they are much too young, being mostly of Mesozoic age. They will be described in chapters dealing with non-orogenic igneous activity (Part III).

4.7.4 Bauxite

As with manganese, these deposits owe their formation to tropical weathering during the Tertiary and Quaternary, but as the host rocks are Birimian greenstones and aluminous phyllites and Eburnian granites, they are dealt with here. Although there must be large deposits in several parts of this extensive terrane, it appears that the only presently exploited bauxite deposits in this age province are in Ghana, where there are plans for considerable expansion of the industry. The open-pit mine at

Awaso has an annual production capacity of 300 000 tonnes. It has large reserves of high-grade ore, and production in 1980 was nearly 250 000 tonnes, most of it exported, though this fell to less than 200 000 tonnes in 1981, and by 1983 it was down to only 70 000 tonnes. Major deposits at Kibi, north-west of Accra, are also exploited, and there is a smelter at Tema in the Volta basin, with a capacity of 200 000 tonnes per year. It is intended ultimately to produce as much as 500 000 to 800 000 tonnes of alumina per year from bauxite to supply the smelter, to which a rolling mill will eventually be added.

4.7.5 Iron ores
Although on a global scale Proterozoic iron formations are generally thicker and more extensive than those of the Archaean, iron ores are a great deal more widespread in the Archaean nucleus of the West Africa craton than in the Lower Proterozoic Birimian terrane, where their place seems to have been taken by manganiferous sediments.

The large Falémé deposit in easternmost Senegal may be of this type, however. It contains some 400×110^6 tonnes of magnetite ore averaging 45–50% Fe, and 100×10^6 tonnes averaging 62–65% Fe. Such a deposit could support an annual production of some 10 million tonnes for 20 years, but this must await construction of links to the main Bamako–Dakar railway as well as preparation of the mine site. There appears to be a continuation of this deposit across the border in southwesternmost Mali, where some 150×10^6 tonnes of haematite ore have been proved, grading between 36 and 67% Fe.

Banded iron formation rocks in southwestern Ivory Coast – for example the Monogaga deposit near Sassandra – have been classed as being of Birimian age, but these deposits occur within the reactivated part of the cratonic nucleus. They are therefore older than Birimian and belong with the Archaean supracrustals.

Deposits of Lower Proterozoic iron ores occur in norites and gabbros in various places. They are probably all magmatic segregations, dominated by titaniferous magnetite, sometimes with significant *vanadium* enrichment, as in northern Upper Volta. Near Takoradi, in southern Ghana, a deposit of such ores is estimated to be 8 km long and a few hundred metres across, and samples have yielded 55% Fe and 12–22% Ti. An iron and steel smelter is under construction at Takoradi and the aim is ultimately to produce some 150 000 tonnes of steel annually, from 1 million tonnes of ore.

4.7.6 Other minerals
Small deposits of other minerals and metalliferous ores are found in many places throughout the Birimian terrane, and only a few of these are shown on Figure 4.3. There is a substantial deposit of lateritic *nickel* and *cobalt* near Koudougou in Upper Volta, estimated at between 30 and 70 million tonnes grading at 1.5% Ni and 0.1–0.3% Co. Similar laterites occur in Ivory Coast and in southwestern Niger, but their size and extent is not known. The Birimian greenstones contain various combinations of *copper*, *lead*, *zinc*, *antimony* and *silver* in many places, for example, near Poura and the Liptako region of Upper Volta, in southwestern Niger, near Kenieba in southwestern Mali, and in Ghana. Chrysotile *asbestos* occurs in serpentinites among the greenstones of southeastern Ghana. *Molybdenite* has been found in Ghana, Upper Volta and southwestern Niger, probably associated with Eburnian granites. Pegmatites associated with granites also yield *tin*, *niobium*, *tantalum*, *beryllium* and *lithium* in various proportions, for example in southern Upper Volta, south-east of Bamako in southwestern Mali and near Tera in southwestern Niger. Lithium-bearing (spodumene) pegmatites also occur near Saltpond, about 90 km west of Accra. Tin was mined in Ghana during World War II and tantalite, assaying 60% Ta_2O_3 in places, has been extracted at Issia in Ivory Coast, from placer deposits derived from pegmatite. *Uranium* showings have been found close to Kedougou in eastern Senegal, though these may be in younger sediments. *Marble* deposits have been found in some places, the one at Tiara in Upper Volta being reported to contain some 60 000 m^3 of workable reserves.

Placer deposits of *rutile* have been exploited along the coast west of Abidjan in Ivory Coast (Grand-Lahou), presumably derived from mafic and/or pegmatitic rocks in the Birimian, but otherwise generally similar to those occurring in Sierra Leone (Sec. 3.9).

As in other Precambrian areas, there is an abundance of materials for constructional and related purposes, including aggregate and crushed rock from granite and migmatite and pegmatitic feldspars (including kaolinised pegmatites, e.g. near Saltpond in Ghana) for ceramics. In general, however, as the areal proportion of Birimian supracrustals far exceeds that of the other two main age provinces in West Africa, the scope for discovery of valuable minerals is correspondingly greater.

49

Bibliography

Amaoko-Mensah, A. 1972. *Geochemical aspects of spodumene pegmatites of the Saltpond area, Ghana*. 16th Ann. Rep., Res. Inst. Afr. Geol., Univ. Leeds, 58–61.

Angoran, Y. and E. Kadio 1983. Aperçu de precambrien de Côte d'Ivoire: géologie–metallogenie. *J. Afr. Earth Sci.* **1**, 167–76.

Attoh, K. 1982. Structure, gravity models and stratigraphy of an early Proterozoic volcanic–sedimentary belt in northeastern Ghana. *Precambrian Res.* **18**, 275–90.

Bard, J. P. 1975. Classification et origines des granitoids éburnéens du craton Ouest-Africain. *C.R. Acad. Sci. Paris* **281**, 867–70.

Bard, J. P. 1976. Evolution géotectonique du craton Ouest-Africain en Côte d'Ivoire; éléments d'un nouveau schéma. In *African geology* (Proc. 2nd Conf. African Geology, Addis Ababa, 1973), Tsegaye Hailu (ed.). Ibadan: Geol. Soc. Africa.

Bard, J. P. and S. Lémoine 1976. Phases tectoniques superposées dans les métasédiments Précambriens du domaine côtier occidental de la Côte d'Ivoire. *Precambrian Res.* **3**, 209–29.

Behrendt, J. C. and C. S. Wotorson 1970. Aeromagnetic and gravity investigations of the coastal area and continental shelf of Liberia, West Africa, and their relation to continental drift. *Geol. Soc. Am. Bull.* **81**, 3563–74.

Béssoles, B. 1977. *Géologie de l'Afrique: le craton Ouest Africain*. Mem. BRGM Fr., no. 88.

Black, R. 1980. Precambrian of West Africa. *Episodes* **1980**, no. 4 (Dec), 3–8.

Brunnschweiler, R. O., A. N. Dempster and I. Kusnir 1972. Precambrian systems in western Niger. In *African geology*, T. F. J. Dessauvagie and A. J. Whiteman (eds). Ibadan: Ibadan University Press.

Burke, K. C. and J. F. Dewey 1972. Orogeny in Africa. In *African geology*, T. F. J. Dessauvagie and A. J. Whiteman (eds). Ibadan: Ibadan University Press.

Cahen, L. and N. J. Snelling 1984. *Geochronology and evolution of Africa*. Oxford: Oxford University Press.

Casanova, R. and I. Yace 1976. Pétrogenèse des orthométamorphites éburnéennes (Précambrien moyen) du centre de la Côte d'Ivoire (Afrique de l'Ouest). *C.R. Acad. Sci. Paris D* **282**, 1241–4.

Dixon, C. J. 1979. *Atlas of economic mineral deposits*. London: Chapman and Hall.

Gole, M. J. and C. Klein 1981. Banded iron-formations through much of Precambrian time. *J. Geol.* **89**, 169–83.

Grandin, G. and E. A. Perseil 1983. Les minéralisations manganésifères volcano-sedimentaires du Blafa-Guéto (Côte d'Ivoire) – paragénèses – altération climatique. *Min. Dep.* **18**, 99–111.

Hastings, D. A. 1982. On the tectonics and metallogenesis of West Africa: a model incorporating new geophysical data. *Geoexploration* **20**, 295–327.

Kesse, G. O. 1976. New concepts on the stratigraphy of the Birimian rocks of Ghana. In *African geology*, Tsegaye Hailu (ed.). Ibadan: Geol. Soc. Africa.

Kolbe, P., W. H. Pinson, J. M. Saul and E. W. Miller 1976. Rb–Sr study on country rocks of the Bosumtwi Crater, Ghana. *Geochim. Cosmochim. Acta* **31**, 869–75.

Lewry, J. F. 1981. Lower Proterozoic arc–microcontinent collisional tectonics in the western Churchill Province. *Nature* **294**, 69–72.

Ntiamoh-Agyakwa 1979. Relationship between gold and manganese mineralization in the Birimian of Ghana, West Africa. *Geol. Mag.* **106**, 345–52.

Olade, M. A. 1980. Precambrian metallogeny in West Africa. *Geol. Rdsch.* **69**, 411–28.

Peron, C. 1975. *Atlas des indices minéraux de la Côte d'Ivoire au 1/4 000 000*. Abidjan: SODEMI.

Perseil, E. A. and G. Grandin 1978. Evolution minéralogique du manganèse dans trois gisements d'Afrique de l'Ouest: Mokta, Tambao, Nsuta. *Mineralium Deposita* **13**, 295–311.

Prasad, G. 1983. A review of the early Tertiary bauxite event in South America, Africa and India. *J. Afr. Earth Sci.* **1**, 305–14.

Sestini, G. 1973. Sedimentology of a palaeoplacer: the gold-bearing Tarkwaian of Ghana. In *Sedimentary ores*, G. C. Amstutz and A. J. Bernard (eds). Berlin: Springer Verlag.

Tempier, P. and S. Lémoine 1969. Données sur quelques dolérites du Sud-est de la Côte d'Ivoire et leurs rapports avec un granite alcalin. *Bull. Soc. Géol. Fr.* (7th ser.) **XI**, 704–9.

Tougarinov, A. I., K. G. Knorre, L. L. Shanin and L. N. Prokofieva 1968. The geochronology of some Precambrian rocks of southern West Africa. *Can. J. Earth Sci.* **5**, 639–42.

UNESCO 1968. *Proceedings of the symposium on the granites of West Africa (1965)*. Natural Resources Research Series, no. VIII. Paris: UNESCO.

Vachette, M. and O. F. Ouedraogo 1978. Age birrimiens déterminés par la méthode au strontium sur les granitoides de la région de Boulsa (Centre-Est de la Haute-Volta). *C.R. Somm. Soc. Géol. Fr.* **4**, 201–5.

Vachette, M., J. M. Cantagrel and P. E. Gamsoure 1975. Ages birrimiens déterminés par les méthodes au strontium et a l'argon sur les formations cristallines et crystallophylliennes de la région de Ouahogpuya (NW Haute-Volta). *C.R. Acad. Sci. Paris, D* **280**, 1329–32.

5 *The Pan African of West Africa – the western domain*

STUDY NOTE

The Pan African orogeny or thermotectonic event is represented in West Africa by the relatively narrow Rokelide belt and the southern part of the Mauritanide belt in the west, and by the extensive Togo–Benin–Nigeria 'swell' in the east. The two domains are separated by the West African craton (Fig. 1.4) and for convenience they are treated in separate chapters.

SUMMARY

In the western Pan African domain, the Rokelides encompass a single low-grade ensialic supracrustal belt on the western margin of the Archaean cratonic nucleus, which suffered slight reactivation during the Pan African deformation and metamorphism of the belt. The Rokelides merge northwards with the Mauritanides. The Rokelide belt is believed to have been a rifted aulacogen that was deformed in the first of a series of plate collision events that culminated in the building of the Mauritanides. As yet, few mineral deposits of importance have been discovered in this part of the Pan African terrane in West Africa.

5.1 Introduction

The term Pan African refers to the major and widespread orogenic or thermotectonic event that affected most of the rocks outside the cratons, between about 650 and 450 Ma ago (Fig. 1.3). It involved the last major reactivation of basement rocks in Africa, and was the final stage in the formation of the African shield. After it, the so-called mobile belts (younger orogens) became as tectonically stable as the cratons themselves. The sole exceptions were the northern parts of the Mauritanides and the Atlas mountains in the northwest, and the Cape Fold belt in the extreme south. Apart from these comparatively small areas, the only deformation to affect the continent following the Pan African event was faulting and gentle (epeirogenic) crustal warping.

5.2 The Rokelide–Mauritanide belt

Only the northern and southern segments of the Rokelide belt are exposed (Fig. 5.1), the middle portion being obscured by the essentially flat-lying Palaeozoic platform sediments of the Bové Basin (the southwestern extension of the larger Taoudeni Basin). The northern segment merges with the Mauritanide belt, in which the effects of later Hercynian deformation and metamorphism are recognised (Sec. 5.2.1).

The sediments and volcanics of the Rokelide belt have been studied in most detail in Sierra Leone, where they are collectively termed the **Rokel River Group**. They range in age from uppermost Precambrian (Infracambrian) to lowermost Phanerozoic, reach a total thickness of about 5 km, and occupy a broadly synclinorial structure within the Archaean basement on the western edge of the cratonic nucleus.

The sediments at the base of the Rokel River Group comprise conglomerates, feldspathic sandstones and clays, which are interpreted as glacial and fluvioglacial in origin, recording advances and retreats of an ice sheet that moved over the region from the east. This is consistent with the evidence of a widespread Infracambrian glaciation in several parts of West Africa, which is further discussed in Part II.

The subsequent history of sedimentation in the Rokelide belt was one of progressive deepening of an elongate depositional trough, which then gradually silted up as delta fronts advanced into it, supplied from the cratonic hinterland to the east. Volcanics near the top of the succession are dominated by andesites, but include spilites with pillow structures.

51

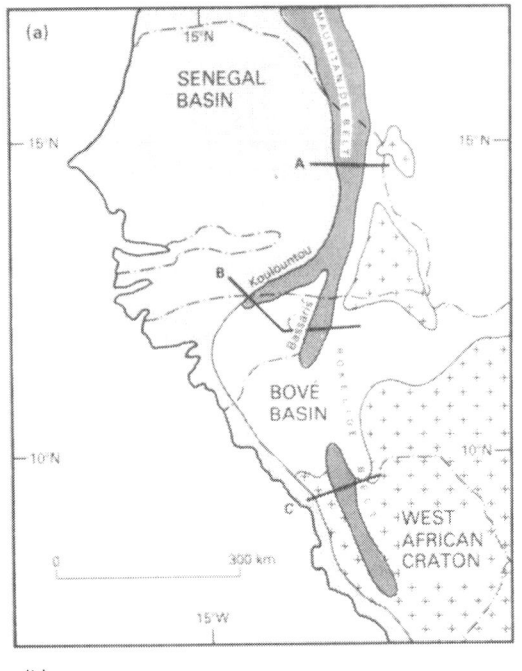

(b)

Metamorphic grades increase westwards across the Rokel River Group, from non-metamorphic lithologies in the east to greenschist facies in the west. Structures in the Rokel River Group are quite different from those that characterise most other supracrustal belts in West Africa, with their mainly isoclinal folding and steeply dipping foliation. Basement reactivation along this belt did not involve large-scale remobilisation and plastic deformation. Here the Pan African orogeny reactivated only brittle structures in the long-stabilised cratonic basement, i.e. it caused movement along ancient faults and zones of weakness. Differential movements of fault-bounded blocks of the underlying Archaean folded the relatively thin cover of sediments draped over them. These differential movements were the result mainly of thrusting from the south-west. On a regional scale, the strike is NNW–SSE, defining the overall trend of this segment of the Rokelide belt. Dips average 20° on the eastern margin, 60°E on the western margin.

Because of the strong fault-controlled nature of the deformation, folds are not uniformly developed and there are zones of intense deformation separated by virtually undeformed rocks. Folds in the

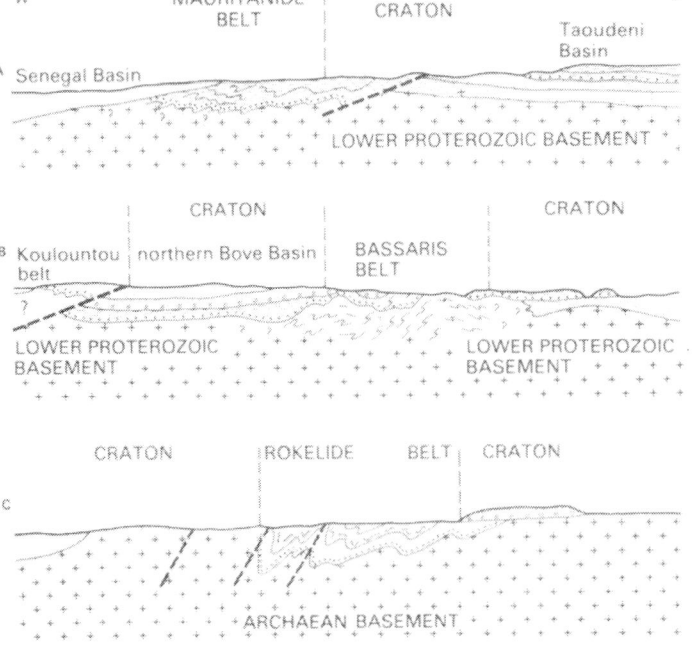

Figure 5.1 (a) Generalised map of the Rokelide–Mauritanide belt of West Africa. Lines A to C refer to cross sections in (b). (b) Diagrammatic cross sections along the lines shown in (a). The basal Infracambrian and basal Ordovician tillites will be discussed in Part II.

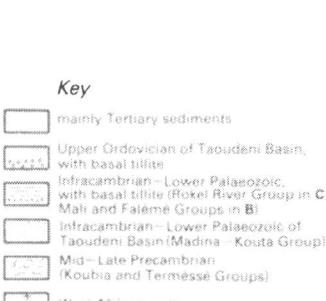

52

deformed zones are asymmetric with southwesterly dipping axial planes and their axes plunge generally south, although the belt as a whole plunges north, beneath the younger sediments of the Bové basin.

The Rokelide belt dies out in southern Sierra Leone, but the Gibi Mountain Formation in north-western Liberia has been tentatively correlated with sediments at the base of the Rokel River Group. It occurs as an outlier upon the basement and is itself overlain by a klippe (an outlying thrust remnant) of quartzite.

The northern segment of the Rokelide belt, in northwestern Guinea and southeastern Senegal, is known locally as the **Bassaris belt** (Fig. 5.1). It is a NNE-trending synclinorial structure, in which the sediments and volcanics of the **Mali Group** can be confidently equated with the Rokel River Group of Sierra Leone. Largely undeformed and unmetamorphosed equivalents of the Mali Group occur on the craton to the east of the Bassaris belt, where they have been called the **Faléme Group**. Both the Mali and Faléme Group sediments are characterised by basal conglomerates and sandstones that have been interpreted as Infracambrian tillites and other sediments of glacial origin. They are not to be confused with the later (Ordovician) tillites and related glacigenic deposits that form part of the post-Pan African sedimentary successions in this region (Part II).

The Mali Group consists mainly of shales, dolomites and quartzites, which predominate over lavas and greywackes. The depositional character is that of a passive continental margin, and the lithologies suggest more open marine conditions than prevailed during the same period further south, where the Rokel River Group was being deposited.

Unlike the Rokel River Group, however, the Mali Group is underlain by older rocks, comprising the **Koubia Group** and **Termésse Group**, eugeosynclinal rocks with abundant volcanics and greywackes and calc-alkaline intrusives, which represent an earlier active continental margin environment and which appear to have been mildly deformed and metamorphosed prior to deposition of the Mali Group. Although these rocks cannot be directly traced eastwards, they may well be lateral equivalents of undeformed Upper Proterozoic sediments that rest upon the craton beneath the Faléme Group and are known as the **Madina–Kouta Group** (Fig. 5.1).

In short, in this region there are two deformed and metamorphosed continental margin successions that were deposited near the western boundary of the West African craton, each with undeformed lateral equivalents upon the craton itself. The two sequences are separated by the Infracambrian glacial deposits that appear to be widespread in this part of West Africa (Part II).

As they lie upon the craton, the Faléme and Madina–Kouta Groups belong to the Taoudeni Basin succession (Part II). Their relationship to the Mali Group and Koubia and Termésse Groups respectively, further west, is very similar to that between the rocks of the Togo belt and the Volta Basin on the eastern edge of the craton (Sec. 6.2). Similar relationships characterise the southwestern branch of the Mauritanides, the **Koulountou belt**. Here there are deformed sediments and volcanics, probably similar to those of the Mali Group, with undeformed equivalents on the cratonic block that separates the Koulountou and Bassaris belts and underlies the northern part of the Bové Basin (Fig. 5.1, line B).

The Mauritanide belt further north in Senegal is dominated by rocks that have been referred to as the **Bakel Group** and are generally similar to those of the Mali Group. Undeformed lateral equivalents of these rocks lie upon the craton (Fig. 5.1, line A) and also form part of the western Taoudeni Basin succession (Part II).

Metamorphic grades in the belts north of the Bové Basin are generally higher than in the Rokel River Group, ranging from greenschist to amphibolite. Structures are more complex in the Bassaris belt than in the southern segment of the Rokelide belt, for the Mali Group is underlain by previously deformed older Proterozoic rocks. But, as this is also an ensialic belt on the craton, deformation was at least partly fault-controlled, as it was further south. In the Mauritanide–Koulountou belt, there is abundant evidence of thrust and nappe tectonics incorporating slices of basement and of ultramafic rocks (some of which, at least, are probably ophiolite fragments), and at least three major episodes of folding have been identified.

Gravity and seismic data are consistent with the relationships suggested by line A of Fig. 5.1. The results of a geophysical traverse across the Mauritanides just north of the Bassaris and Koulountou branches indicate that the craton margin at depth lies some 80 km west of where it outcrops on the surface. The more strongly deformed sediments and volcanics in the west of the belt that were thrust upon the craton margin correspond to the positive gravity anomaly in Fig. 5.2. The less strongly deformed rocks in the east of the belt and overlying the craton coincide with the negative gravity anomaly. Similar profiles have been determined in other parts of the Mauritanide belt, and it should be noted that there is a linear gravity anomaly along the Rokelides, which continues right across beneath the Bové Basin.

5.2.1 Geochronology in the Rokelide–Mauritanide belt

The lowermost strata of the Rokel River Group are believed to be of Infracambrian age (c. 620 Ma). There are no direct geochronological data for the Rokel River Group rocks themselves, but the effects of the Pan African event on the surrounding Archaean basement are recorded in K/Ar ages which date the event at c. 560 Ma. It is important to recall that the Kasila Group and the Marampa Group have also been affected by the Pan African event, and indeed the majority of ages recorded from these belts are Pan African (which has led to the suggestion that the Kasila and Marampa Groups may be Pan African belts, as noted in Section 3.6).

Further north, in Guinea and Senegal, the cratonic basement is Birimian (Fig. 5.1), formed of isoclinally folded pelitic sediments, greywackes, jaspers and conglomerates, as well as a variety of igneous rocks, including pillow lavas and ultramafic lithologies, all metamorphosed in the greenschist facies. Granitic pebbles in the conglomerates indicate derivation from a pre-Birimian basement, and granitic intrusions give Eburnian ages (2010–2090 Ma).

The prevailing view of age relationships in the Bassaris and Mauritanide belts (i.e. north of the Bové Basin, Fig. 5.1) is that the Pan African event is difficult to recognise because of later tectonism. An episode of deformation dated at c. 435 Ma is believed to have folded and metamorphosed the Mali Group in Guinea and Senegal, effectively obliterating the effects of the Pan African. Further north still, the Hercynian orogeny (c. 355 Ma) is held responsible for most of the deformation and metamorphism of the Bakel Group. In other words, the effect of the Pan African event dies out northwards, to be replaced first by the c. 435 Ma event, then by the Hercynian orogeny, which appears to have been the principal cause of deformation and metamorphism in the Mauritanide–Koulountou belt (cf. Figs 1.3 and 2.1). At first sight the distribution of these ages implies that the Mauritanides were

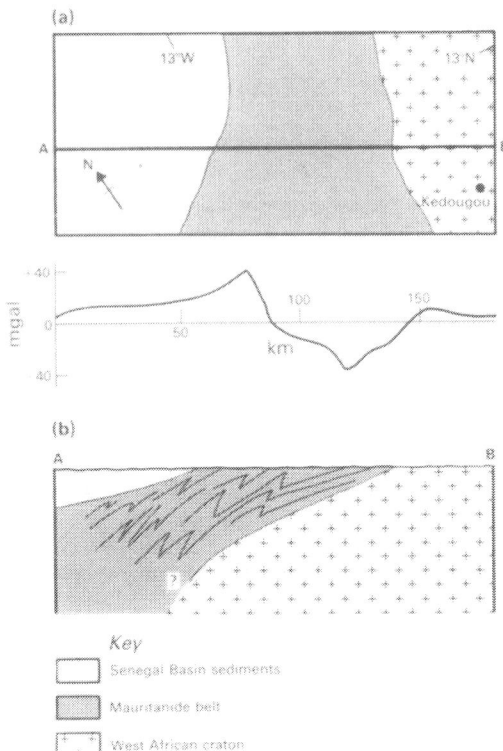

Figure 5.2 (a) Bouguer gravity anomaly profile across the Mauritanide belt. (b) Schematic cross section (not to scale) showing the likely geological relationships.

subjected to a strongly diachronous deformation, which became progressively younger northwards – from Pan African in the south to Hercynian in the north. A more plausible interpretation is that the Mauritanide–Koulountou belt experienced successive deformations which were of different intensity in different parts of the belt.

5.2.2 Plate tectonics and the Rokelide–Mauritanide belt

The presumed ensialic setting of the Rokelide belt (Fig. 5.1) means that it could not have formed directly by the collision of continental plates as illustrated in Figure 2.2. It could have formed by thinning and rifting of continental crust due to back-arc spreading above a subduction zone (cf. Fig. 2.6). Subduction would have been directed eastwards, however, leading to considerable Pan African reactivation of West African craton. Such reactivation is recorded only along the craton margin and is restricted to relatively local thermal effects, involving minor retrograde metamorphism and the intrusion of a few granites.

Alternatively, the Rokelide–Bassaris belt may be a 'failed-arm' or **aulacogen**, an intracontinental rift branching inland from a passive continental margin now represented by the Mauritanide–Koulountou belt. Deposition of cherts, dolomites, greywackes and volcanics occurred along this margin and in the northern part of the rift, where the Mali Group was deposited. Further south, shallower-water conditions prevailed, with the deposition of the Rokel River Group. The contrasted character of the Mali Group and the Rokel River Group suggests that the rift was wider at the northern end and remained open marine in character, while in the south a cycle of deepening and shallowing water conditions occurred as the rift formed and filled. The calc-alkaline character of the volcanics (mainly andesites) in the Rokel River Group is not wholly consistent with this simple picture, for such rocks normally form above subduction zones rather than in rifted regions. As outlined above, however, subduction would have to be directed eastwards, and this seems unlikely.

Deformation and metamorphism of the Rokel River and Mali Groups can be attributed to a continental collision that built a Pan African precursor of the Mauritanide–Koulountou belt, rejuvenating earlier faults and thrusts on the craton to the southeast.

The main Hercynian deformation of the Mauritanides must also have involved continental collision, for this portion of the belt is characterised by the development of major thrusts and nappes. This in turn implies a Palaeozoic episode of oceanic opening and closure, after the Pan African event. The effects of the Hercynian orogeny were much stronger in the northern Mauritanides than in the south, and there is palaeomagnetic evidence of transcurrent movements in the deformation, implying that the collision may have been oblique rather than 'head-on'.

This brief review cannot be the whole story. It ignores the probability that two major deformations affected the Bassaris belt, one of which is presumably to be correlated with the 'intermediate' c. 435 Ma (Caledonian) age recorded for this region. The Pan African (c. 550 Ma) and Hercynian (c. 350 Ma) events seem to be reasonably well documented for the southern and northern parts of the Rokelide–Mauritanide belt respectively. The full history of the middle segments remains to be worked out, but

it is worth noting that the events reviewed here must be closely related to the evolution and closure of the Palaeozoic proto-Atlantic Ocean.

5.3 Economic potential of the western Pan African domain

Relatively few mineral deposits of importance appear so far to have been discovered in the Rokelide–Mauritanide belt of southern West Africa, but this is not to say that more do not exist. Those that have been found occur in and near eastern Senegal and include *copper* near Bakel, *uranium* near Kedougou and *iron ore* (magnetite) near Goto. In addition, at least 350 000 tonnes of high-grade marble have been proved near Ibel and production of this material is planned eventually to reach 30 000 tonnes annually.

Bibliography

Allen, P. M. 1980. Discussion of Culver and Williams 1979. *J. Geol. Soc. Lond.* **37**, 511–12.

Bassot, J. P. and M. Caen-Vachette 1983. Données nouvelles sur l'âge du massif de granitoide du Niokolo-Koba (Sénégal Oriental); implications sur l'âge du stade précoce de la chaine des Mauritanides. *J. Afr. Earth Sci.* **1**, 159–66.

Black, R. 1980. Precambrian of West Africa. *Episodes* **1980**, no. 4 (Dec.), 3–8.

Briden, J. C., D. N. Whitcombe, G. W. Stuart, J. D. Fairhead, C. Dorbath and L. Dorbath 1981. Depth of geological contrast across the West African craton margin. *Nature* **292**, 123–8.

Cahen, L. and N. J. Snelling 1984. *Geochronology and evolution of Africa.* Oxford: Oxford University Press.

Culver, S. J. and H. R. Williams 1979. Late Precambrian and Phanerozoic geology of Sierra Leone. *J. Geol. Soc. Lond.* **136**, 605–18.

Culver, S. J., H. R. Williams and P. A. Bull 1978. Infracambrian glaciogenic sediments from Sierra Leone. *Nature* **274**, 49–51.

Gerard, A. and M. Ogier 1979. L'étude des gisements de magnetite par prospection géophysique: cas particulier de l'étude gravimétrique de gisement de Goto (Sénégal oriental). Chron. Rech. Min., BRGM, no. 447, pp. 5–26.

Grant, N. K. 1973. Orogeny and reactivation to the west and southeast of the West African craton. In *The ocean basins and margins*, vol. 1, A. E. M. Nairn and F. G. Stehli (eds). New York: Plenum.

Williams, H. R. and S. J. Culver 1982. The Rokelides of West Africa – Pan African aulacogen or back-arc basin. *Precambrian Res.* **18**, 261–74.

6 The Pan African of West Africa – the eastern domain

SUMMARY

East of the craton there are many low-grade supracrustal belts mainly occupying a broad elongate NNE–SSW zone in the western half of Nigeria. They are dominated by clastic metasediments, but basic and ultrabasic and banded iron formation lithologies also occur. Most of the belts are probably late Proterozoic in age, deformed and metamorphosed in the Pan African (550 Ma), though some bear the imprint of a Kibaran event (c. 1100 Ma). The extensive basement to the supracrustals has a record of reactivation extending back at least to the Liberian and it contains many scattered older supracrustal relics. Metamorphic grades in the basement are in the amphibolite facies throughout most of the region, except near the margin of the craton, where granulite facies rocks are common in Ghana, Togo and Benin. There are abundant large syntectonic to late-tectonic granitic intrusions, most of which give Pan African ages (c. 450–650 Ma). The eastern boundary between the craton and the Pan African terrane is a broad zone of high-angle thrusting towards the west, called the Togo belt. This boundary has been identified as a crustal suture formed by collision with a somewhat poorly defined East Saharan craton lying to the east of it. The Dahomeyan terrane east of the thrust zone may have formed by aggregation of smaller continental fragments, island arcs and intervening basins. The case for identifying the Pan African event in the eastern part of West Africa as a collision orogeny is strengthened by well documented evidence from the Hoggar region of Mali and southern Algeria.

The best-known economic mineral deposits in the eastern Pan African domain are of gold in supracrustal belts and of tin–tantalum–niobium minerals in pegmatites. There are many small deposits of other minerals and at least one major occurrence of iron ore.

6.1 Introduction

In this large eastern Pan African domain, there are many low-grade supracrustal belts whose size and general NNE–SSW trend is similar to that of supracrustals in the cratonic nucleus (Fig. 2.4). However, they are largely confined to a broad belt in the western half of Nigeria, except for scattered outlying ridges of mainly quartzitic rocks nearer to the craton margin, in Benin and southern Togo. The basement has a history of reactivation going back at least to the Liberian and it experienced its last major reactivation in the Pan African. The supracrustals have been strongly deformed, being almost everywhere isoclinally folded with a steep foliation that parallels the trend of the belts. Metamorphism is generally in the greenschist to amphibolite facies.

A large area on the eastern side of the craton is overlain by mainly flat-lying sediments of late Proterozoic (Infracambrian) to early Palaeozoic age that occupy the Volta Basin (Fig. 1.4). This major sedimentary basin demonstrates the important distinction between stabilised and reactivated crust. Most of the platform sediments are contemporaneous with deformed and metamorphosed supracrustals of the Buem and Togo Formations. These are thought to lie on Dahomeyan basement in and near the thrust zone which forms the margin of the West African craton (the Togo belt).

6.2 The Togo belt

The **Togo belt** comprises supracrustal sediments and volcanics of probable late Precambrian to early Phanerozoic age, deformed and partly metamorphosed by powerful northwesterly directed thrusting of Dahomeyan basement rocks onto the West African craton. Their undeformed lateral equivalents in the Volta Basin (the Voltaian) will be considered in more detail in Part II. The rocks of the Togo belt are also widely held to be of the same age (i.e. Katangan, Table 2.1) as most of the supracrustal belts further east in the main part of the Pan African domain (Fig. 6.1).

Closest to the craton is the **Buem Formation**, also called the **Thiélé Unit**, which forms a band of generally flat country about 15 km across on average,

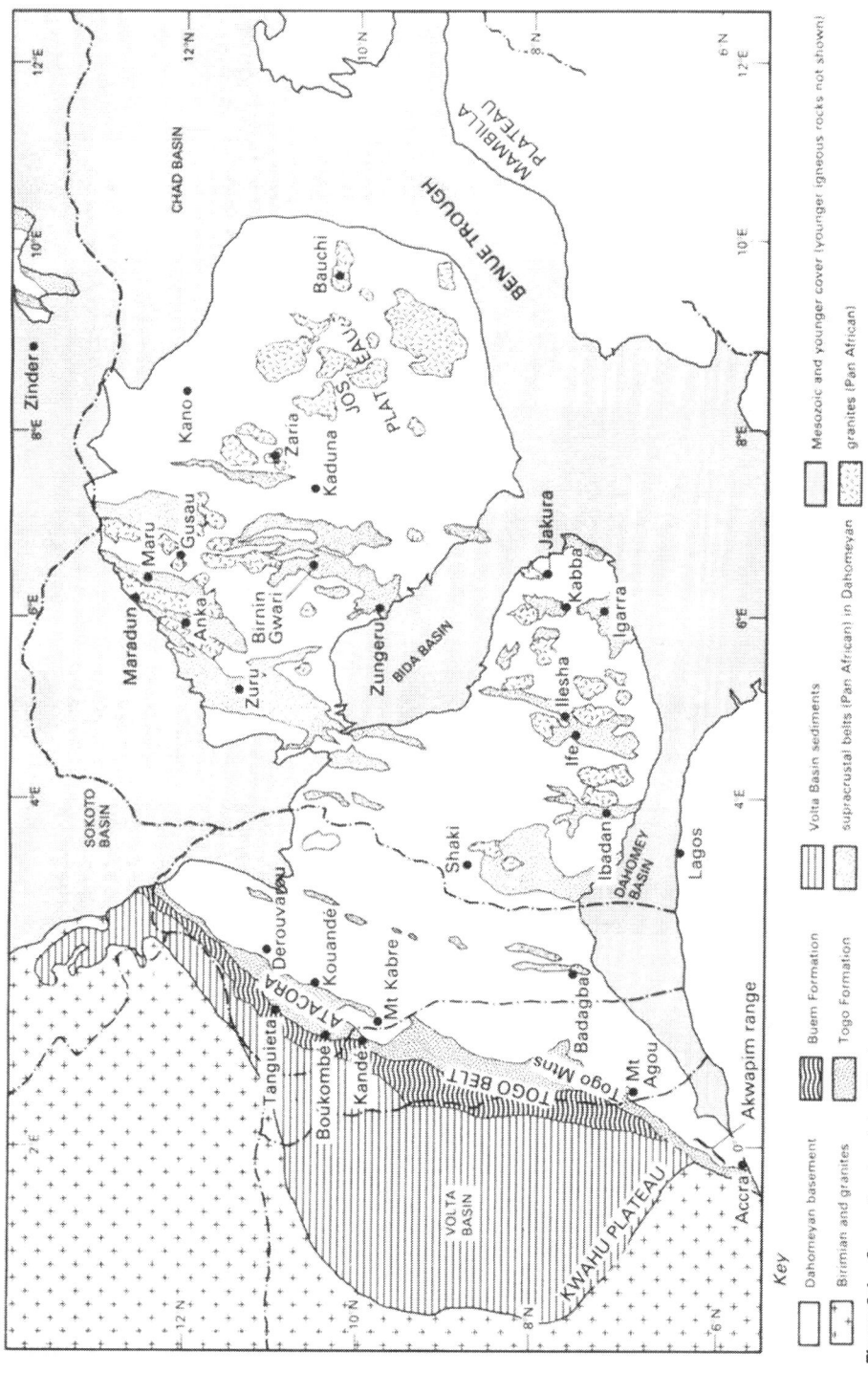

Figure 6.1 Generalised and simplified map of the main part of the Pan African of the eastern domain of West Africa (Younger igneous rocks are not shown)

Key

☐ Dahomeyan basement

☐ Birimian and granites

▦ Buem Formation

▦ Togo Formation

▦ Volta Basin sediments

▦ supracrustal belts (Pan African) in Dahomeyan

☐ Mesozoic and younger cover (younger igneous rocks not shown)

▦ granites (Pan African)

with scattered small hills. It defines the eastern limit of the Volta Basin. The rocks constitute a southeastward dipping sequence dominated by clastic sediments, mainly sandstones and siltstones, shales and mudstones being subordinate. There are some massive cherts (**silexites** of French writers), also limestones, dolomites and sedimentary ironstones. Conglomeratic horizons near the base of the succession have been interpreted as glacial tillites. Volcanics interstratified with sediments in the succession include rocks of both alkaline and calc-alkaline affinities. At least some of them were erupted under water, for pillow structures are preserved. In many places throughout the belt, schistose and massive serpentinites, some of them chromite-bearing, have been tectonically emplaced along the numerous thrust planes that cut the succession. They probably represent slices of upper mantle. There are also cross-cutting dolerites, which may be of late or even post-Pan African age (Sec. 6.3.3).

Rocks of the Buem Formation are largely unmetamorphosed. Deformation is mainly the result of thrust tectonics with much **imbrication**, the stacking of inclined thrust sheets one upon the other, and the overall dip is to the south-east. Two generations of thrust movements have been recognised. Mylonites and cataclastic rocks are common and the thrust zones are often marked by brecciated silexites and serpentinites.

There has been much repetition of the succession by the thrusting, and this has combined with generally poor exposures to prevent unequivocal elucidation of the structures. In Ghana, for example, it has been commonly supposed that the Buem Formation is strongly folded into asymmetric overturned structures with southeasterly dipping axial planes. This interpretation places the volcanics near the top of the succession and yields estimates for the total thickness of these largely unmetamorphosed rocks of around 3600 m. An alternative view is that the appearance of strong folding is illusory, due to pinching and swelling of sandstone horizons resulting from rapid lateral facies variations. Where younging directions can be determined the beds are the right way up, and although small chevron folds with steep axial planes are common in fine-grained beds, massive sandstones are not folded. This interpretation would place the volcanic rocks towards the base of the succession rather than at the top, and is in accord with the findings of geologists in Togo and Benin. It also leads to a maximum possible figure for the thickness of the Buem Formation of 40 000 m,

including 5000 m of volcanics, but this makes no allowance for repetition of the sequence by thrusting, and the true thickness must be very much less than this.

The **Togo Formation** immediately to the east of the Buem Formation is called the **Akwapimian** in Ghana and the **Atacorian** or **Atacora Unit** in Togo and Benin. It occupies an irregular 5–50 km wide strip bordered on the west by thrust contacts against the Buem Formation, on the east by thrust contacts with the Dahomeyan basement. It includes the Atacora range in Benin, the Togo Mountains and the Akwapim range in southern Ghana, where the Buem Formation wedges out and the Togo Formation defines the eastern boundary of the Volta Basin (Fig. 6.1).

There are two principal lithologies, one psammitic, the other pelitic. Quartzitic sandstones and quartzites contain conglomeratic layers, and some quartzites are ferruginous. Phyllites and mica-schists, including the **Kandé–Boukombé Series**, also have conglomerate horizons, as well as basaltic volcanics, now metamorphosed to greenschists. Marble is recorded north of Boukombé. Tectonically emplaced slices of basement rocks are mainly gneisses but include eclogitic and granulite facies rocks of the nearby high-grade Dahomeyan (Sec. 6.3.1), as well as elongate lenses and pods of serpentinite.

The Togo Formation rocks are not only more highly metamorphosed than those of the Buem Formation, they are also considerably more deformed. Both metamorphism and deformation increase towards the south-east. Minor folds are isoclinal with axial planes that dip to the south-east and are generally parallel to lithological layering. Major folds are upright, with only slight tendency to a southeasterly dip. The thrust movements have been correlated with these folds, though the thrust planes appear themselves to have been folded by a later phase of deformation.

The relative age of the Buem and Togo Formations remains a matter of speculation and debate. They may be lateral equivalents, but there seems to be growing agreement that the Togo Formation is the older. At first sight this is at variance with the relationships in Figure 6.1, which shows that in this eastward-dipping belt the Togo Formation overlies the Buem, which implies that it should therefore be younger. The Togo Formation was probably unconformable on the Dahomeyan, however, and was brought up from deeper stratigraphic levels by the thrust faulting, to lie upon the younger Buem rocks.

The stratigraphy of the Volta Basin is dealt with in Part II and so details are not shown in Figure 6.1. However, some comments on relationships between it and the Togo belt are appropriate here. In brief, the Togo Formation is correlated with the Lower Voltaian, partly on the grounds of similar lithologies, both groups of rocks being dominated by alternations of sandstones and shales (quartzites and phyllites or schists), with occasional conglomerates and limestones (marbles). In addition, there is evidence that in southeastern Ghana, where the Lower Voltaian of the Kwahu Plateau abuts against the Togo Formation of the Akwapim range, there is a progressive increase in intensity of folding and metamorphism eastwards from Lower Voltaian into Togo Formation rocks. Similar relationships obtain at the northern end of the Togo belt, where the virtually undeformed **Kirtachi Quartzite** represents the Lower Voltaian in southwestern Niger.

The Middle Voltaian lies unconformably on the Lower Voltaian with a basal conglomerate that can be correlated with conglomerates of the Buem Formation. Both sets of conglomerates have been interpreted as tillites, possibly equivalent to the Infracambrian tillites of the Rokelide–Mauritanide belt west of the craton (Sec. 5.2) and in the Taoudeni Basin (Part II). The overlying sediments of the Middle Voltaian include shales, siltstones and sandstones, and limestones, dolomites and cherts, lithologies similar to those of the Buem. There are no volcanic rocks, but among the arenaceous facies are greywackes of possible volcanogenic origin.

The Upper Voltaian has no equivalent in the Togo belt, and it is believed to be a molasse deposit formed by erosion of Buem and Togo Formation rocks, following their deformation and uplift in the Pan African event. Interpretations of gravity data across the Togo belt suggest that the deep structure here approximates to a mirror image of that on the west side of the craton (Fig. 5.2).

6.3 The eastern domain

The eastern domain of the Pan African rocks shown in the eastern part of Figure 1.4 is only a small part of the huge expanse of country that lies between the West African and Congo cratons and is underlain by rocks affected by the Pan African thermotectonic event (Fig. 1.3). Much of this great region probably consists of reactivated older crust, for while Pan African ages (in the 650–450 Ma range) are recorded from rocks throughout it, both Liberian and Eburnian ages have been obtained from several localities. The Pan African event was the last reactivation to affect this whole region, which is now as tectonically stable as the craton itself and forms an integral part of the African shield.

6.3.1 The basement complex

There seem to be no fundamental differences between the basement in this region and in other parts of West Africa. Granulite facies rocks are most abundant close to the margin of the craton, immediately to the east of the Togo belt, in southeastern Ghana and in Togo and Benin. Gneisses with garnet, pyroxene and scapolite occur among more ordinary quartzo-feldspathic biotite and hornblende-bearing varieties. **Eclogites** (high-pressure garnet–pyroxene rocks chemically equivalent to basalt) have been recorded from among large masses of mafic gneisses that include amphibolites and pyroxenites and contain much garnet. The high-grade rocks are generally considered to have been brought up from deeper crustal (and upper mantle) levels by the westward thrust movements that gave rise to the deformation of the Togo belt (Sec. 6.2).

Further east, in Nigeria, granulite facies rocks in the basement are confined to charnockite bodies, which are generally associated with granites and are probably of igneous origin (Sec. 6.3.3). Here the basement is dominated by quartzo-feldspathic biotite- and hornblende-bearing gneisses, schists and migmatites. Metamorphism is generally in the amphibolite facies, as indicated by the occurrence of index minerals such as garnet, sillimanite, kyanite and staurolite in rocks of suitable composition.

Intercalated among the gneisses and migmatites are numerous supracrustal relics. They have been collectively termed the **Older Metasediments** in Nigeria – and they are likely to include remains of supracrustal belts of the Liberian and Eburnian cycles. Most obvious are prominent ridge-forming quartzites, sometimes sillimanite- or kyanite-bearing and often micaceous, grading into muscovite–quartz–schists. They may be structurally complex (Fig. 6.2) and this helps to distinguish them from the more simply folded quartzites of the Younger Metasediments in the low-grade supracrustal belts (Sec. 6.3.2), which are also generally free of sillimanite or kyanite. There is a metamorphosed banded iron formation, rich in magnetite and haematite, south-east of Kabba in southern Nigeria, where it is

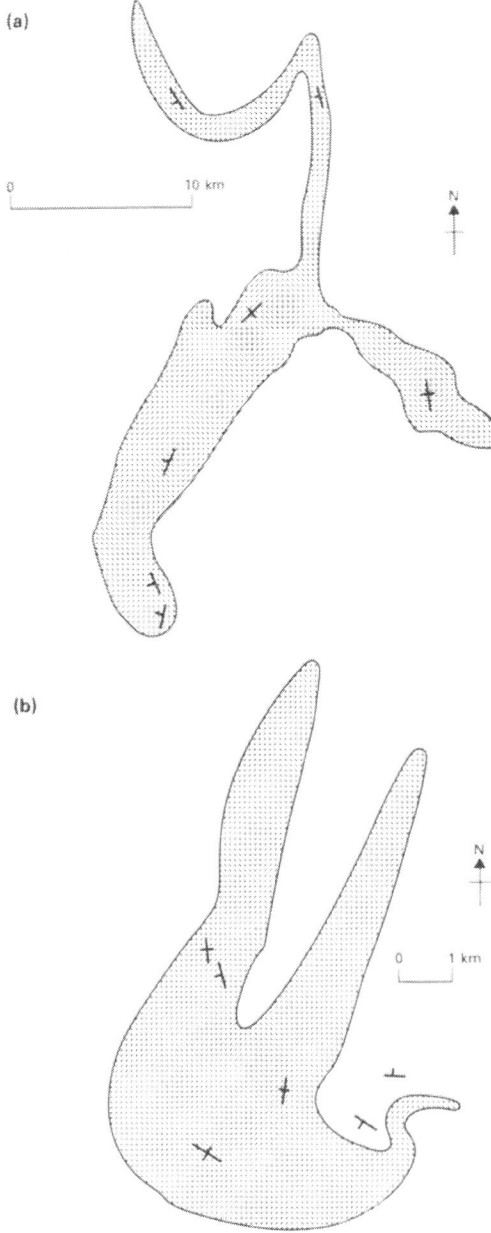

Figure 6.2 Complicated outcrop patterns displayed by quartzites in the Dahomeyan basement of Nigeria, probably disrupted remains of Older Metasedimentary supracrustal belts: (a) north of Zungeru in central Nigeria; (b) east of Zuru in northwestern Nigeria.

an important source of iron ore (Sec. 6.5): smaller occurrences of such rocks are known from the basement in northwestern Nigeria.

Also among the Older Metasediment relics are lenses of more or less dolomitic marble, especially in the vicinity of Kabba and Jakura in southern Nigeria (where some of them are quarried; Sec. 6.5), and numerous small thin sheets and lenses of calc-silicates that are probably metamorphosed impure limestones.

Amphibolites are widespread throughout the basement, ranging in size from small disrupted remnants in the gneisses and migmatites up to larger layers and lenses a few hundred metres long. They probably represent igneous intrusives and volcanics of basic to intermediate composition. Small masses of ultramafic rocks (mostly serpentinite and talc rocks) have also been recorded, at least some of them probably emplaced tectonically along deep fracture zones during basement reactivation.

As in other parts of the West African Precambrian, structures in the basement appear relatively simple on a regional scale. The often steeply dipping foliation is parallel to lithological layering almost everywhere, and the overall structural trend is NNE–SSW over wide areas, parallel to that of the low-grade supracrustal belts (Fig. 6.1).

Local diversions from the regional trend occur in several places, and in eastern parts of Nigeria, especially north of the Benue Trough, the regional trend swings to more nearly ENE–WSW. Poor exposures mean that minor structures are not commonly seen and they are difficult to interpret, especially those produced by plastic flow of semi-molten rock.

6.3.2 The supracrustals

In their overall aspect, the numerous low-grade supracrustal belts are similar to those elsewhere in West Africa: they are synclinorial, the rocks are characterised by tight to isoclinal folding and steeply dipping foliation, and boundaries with the basement (where exposed) are gradational, faulted or sheared. No unconformities have yet been positively identified.

As Figure 6.1 shows, they are mainly confined to a broad NNE–SSW zone in the western half of Nigeria, where they are referred to as the **Younger Metasediments**. This reflects their principal difference from supracrustal belts on the craton: there are few volcanic rocks among them. The Katangan belts are dominated by argillaceous (pelitic) lithologies, represented by phyllites, muscovite–schists and biotite–schists. Quartzites are important in several belts, where they form prominent strike ridges. A few belts have ferruginous and banded quartzites that resemble banded iron formation rocks, and small occurrences of spessartite-bearing quartzite have also been found. Conglomeratic horizons occur in some of the belts, and marbles and calc-silicates are not uncommon.

Other lithologies among the Younger Metasediments represent minor volumes of igneous rocks: amphibolites were probably contemporaneous lavas or minor intrusions, while serpentinites and other ultramafic rocks were probably emplaced tectonically, along deep fractures during deformation of the supracrustal belts. There are also small amounts of acid metavolcanics of dacitic to rhyolitic composition.

Metamorphic grades are variable among the supracrustals. Amphibolite facies rocks are commoner than greenschist facies rocks in the southern belts (Fig. 6.1), where staurolite and almandine–garnet are common constituents of the schists. These index minerals are also found in the northern sector, but more rarely, and most of the supracrustal phyllites here are in the greenschist facies.

Although it is possible to group the supracrustal belts on the basis of common lithologies, such a grouping can have no implications about their relative ages. As on the craton, it is difficult to establish correlations between separate supracrustal belts. The quartzites and mica–schists that form concordant lenses in the basement of Benin (the **Badagba Quartzites**) may be older metasedimentary relics, but it is more likely that they are infolded outliers of the Togo Formation and therefore of much the same age as the Younger Metasediments. They are isoclinally folded and often have strongly sheared contacts with adjacent Dahomeyan rocks.

Major tight to isoclinal folds are often traced out by prominent ridges of resistant rocks among the Younger Metasediments, especially quartzites, and these show up well on aerial photographs. When the closures of such folds are examined in the field, they are invariably obscured by intense shearing parallel to the axial plane foliation. Minor structures are relatively abundant and well exposed in the supracrustal belts, however, and analysis of these has revealed as many as four phases of deformation in some belts.

Evidence of an early phase of thrusting and recumbent folding has been found in parts of Nigeria. It has been all but obliterated by the later NNE–SSW

structures that characterise the supracrustal belts. However, abundant quartzites in part of the most northwesterly belt, near Zuru, are unusual in that they display a predominance of east–west trending structures with shallow dips. These may represent an earlier folding episode in the supracrustal belts, preserved because of the greater structural competence of quartzites compared with that of phyllites and schists.

The **Anka belt** in northwestern Nigeria (Fig. 6.1) stands out from all of the others in having fluviatile sequences of conglomerates, sandstones and siltstones, as well as acid volcanics and minor intrusives that have largely escaped the main effects of Pan African deformation and metamorphism, except along relatively narrow shear zones, where there was tight folding, with stretching and flattening of pebbles. Otherwise, the rocks are mostly only gently folded and sedimentary structures and igneous textures are commonly well preserved in them. It would seem that the deformation was largely fault-controlled as in parts of the Rokelide belt (Sec. 5.2). Relationships with the phyllites which make up most of the Anka belt are not known because of poor exposure. However, the conglomerates contain fragments of phyllite, boulders of quartzite and blocks of granite that are over a metre across.

The late- to post-orogenic volcanic and intrusive acid rocks of the Anka belt are mainly dacitic and rhyolitic lavas and pyroclastics, and dykes and plugs of microtonalite. Particularly striking is a caldera-like structure near Maradun, in which bedded agglomerates and tuffs, some of them waterlaid, have a concentric inward dip. A relatively undeformed and unmetamorphosed volcanosedimentary series comprising spilitic greenstones with pillow structures, acid lavas, sandstones, greywackes and conglomerates with pebbles of basement and volcanic rocks has been described from just to the north-east of Badagba in southern Benin, and appears to be generally similar to the Anka belt.

In the north of Figure 6.1 is the Dahomeyan inlier of Zinder, where basement gneisses and migmatites occur along with supracrustal schists and quartzites. Scattered outcrops of schists and quartzites are also known from the Daura area near the Nigeria–Niger border north of Kano, and these provide evidence of a generally NNE-ward continuation of Younger Metasediments beneath younger sediments of the Chad Basin. South-east of the Benue Trough, the highlands of the Mambilla Plateau extend into

Cameroun and are formed mainly of basement gneisses and granitic intrusions. The northwestern half of Cameroun is part of the Pan African domain of West Africa, while the southeastern half belongs to the Congo craton, which is of Liberian age in this region (Figs 1.3 & 1.4), though there is evidence of both Eburnian and Pan African reactivation along the margin. Supracrustal belts of schists and quartzites in northern and eastern Cameroun and western Chad are assigned by some authorities to the Pan African, by others to the craton.

6.3.3 Granites

There are many syntectonic to late-tectonic intrusions, mainly rocks of granitic composition, but including diorites and syenites. They were emplaced into both basement and supracrustals during or just after the main Pan African deformation. In Nigeria they are called the **Older Granites** to distinguish them from the Jurassic Younger Granites of the Jos Plateau (Part III). They range in size from small subcircular cross-cutting stocks to large elongate concordant batholithic bodies, predominantly of granodioritic composition. A distinctive adamellitic variety with microcline megacrysts up to 5 cm and more in size is called the **Porphyritic Older Granite** in Nigeria. The smaller discordant intrusions are more variable in composition, as they are in the Eburnian terrane. They include the dioritic and syenitic varieties, though most are granites, and there are some unusual types, such as a potash- and magnesia-rich basic syenite associated with pyroxenites, found in south-western Nigeria. Occasional gabbros also occur, one of the largest being in the north-west, south of Zuru. Among these late- to post-tectonic intrusions, there are ring complexes not unlike those that typify the Jurassic Younger Granites (Part III). Good examples are near Toro (granite and diorite) and south of Maru (granite and syenite).

Figure 6.1 suggests that the intrusions are most abundant in the vicinity of the supracrustal belts and in the central part of Nigeria. This is the result of geological mapping, which has historically been concentrated in these regions for economic reasons (Sec. 6.5). More recent mapping is filling some of the gaps.

Charnockites (pyroxene-bearing acid to intermediate igneous rocks) are also associated with Older Granites in some parts of the basement, notably in two main areas: in southern Nigeria, east of Ibadan, and in the north-east, where there is a distinctive fayalite-bearing variety called **bauchite**, named after the type locality, Bauchi. The mineralogy and petrology of these rocks suggest an origin as high-temperature anhydrous melts deep in the crust, prior to formation of the Older Granites with which they are associated.

Late- to post-tectonic dykes of basalt and dolerite have been recorded from many parts of the Dahomeyan basement, in places as far removed as southeastern Ghana and north-east Nigeria. Some bear evidence of incipient metamorphism and deformation and are undoubtedly late Pan African, but radiometric data confirm that there are unaltered dykes of this age as well. However, there are also dykes related to the Younger Granites (Part III) and it is not always possible to tell which is which.

6.3.4 Fractures, faults and mylonites

Major shear zones are marked by discontinuous ridges of mylonitised and silicified rocks and lenses of vein quartz. An early phase of shearing and silicification developed virtually parallel to foliation, most commonly between basement and supracrustal belts. Later transcurrent faulting was mostly at a small angle to foliation, trending between NNE–SSW and NE–SW. Dextral movements amounting to a few tens of kilometres have been recorded in a few places – the best example is near the Anka belt.

Regional studies of fracture systems in the Pan African domain of both southern West Africa and the Hoggar region (Sec. 6.3.6) have revealed a conjugate pattern of NE–SW dextral and NW–SE sinistral faults, which cut earlier mylonitic shear zones. The pattern is consistent with an overall east–west directed stress system (Sec. 6.4).

Two long NNE trending zones of mylonitised basement rocks, several kilometres across, extend north from Zungeru on either side of the Birnin Gwari belt (Fig. 6.1). These are the **Zungeru mylonites**, which are thought to have originated as a major thrust between basement and supracrustals. The mylonites have been recrystallised and folded at least four times, along with the adjacent supracrustals. It must be recorded, however, that these mylonitic rocks were originally mapped as metasediments and metavolcanics, and that some workers still support this identification.

Further south, gently dipping quartzites of the **Effon Psammite Formation** (part of the supracrustal belt near Ilesha, Fig. 6.1) may be in thrust-faulted contact with basement gneisses.

6.3.5 *Correlation and geochronology*

Relict Eburnian and Liberian ages obtained from a number of localities in the Dahomeyan basement, by Rb/Sr and U/Pb methods, provide further evidence of its long history and polycyclic evolution. Apart from these relict ages, most Rb/Sr and all K/Ar age determinations on basement, supracrustals (including those of the Togo belt) and Older Granites, yield Pan African dates in the 650–450 Ma range. The simple conclusion is that deformation and metamorphism of the supracrustals occurred during the Pan African thermotectonic event, which also reactivated large tracts of older continental crust.

The structural and metamorphic uniformity among the supracrustal belts of Younger Metasediments throughout the region supports such an interpretation, allowing for rather higher metamorphic grades in the south of Nigeria compared with the north (Sec. 6.3.2).

However, Kibaran ages (*c.* 1100 Ma) have been obtained by Rb/Sr techniques at a few localities in the basement, and from supracrustals of the Maru belt (Fig. 6.1). At least two Older Granites have given ages of around 750 Ma. There is therefore some geochronological evidence for a Kibaran thermotectonic event affecting the Dahomeyan basement and possibly an older generation of supracrustals. Earlier structural studies in some supracrustal belts led to a similar conclusion, but they have not been fully corroborated by later research. It may also be relevant that marine transgressions into the Volta and Taoudeni Basins commenced about 1000 Ma ago (Ch. 9). These transgressions could have been triggered by an orogenic event east of the craton.

Whatever status is ultimately accorded to the Kibaran event in this domain, however, there is no doubt that it was vastly overshadowed by the Pan African, the waning stages of which appear to have been marked by the intrusion of basalt and dolerite dykes. One of these gave an age of about 480 Ma, which probably dates the time of emplacement, for the rock is unmetamorphosed.

6.3.6 *Correlation with the Hoggar*

North of the area covered in Figure 6.1 the Pan African domain disappears beneath the Mesozoic and younger sediments of the Iullmedden and Chad Basins. It reappears nearly a thousand kilometres to the north, in the Hoggar massif of southern Algeria, with its two southward projections, the Aïr of north-

ern Niger and the Adrar des Iforas of northeastern Mali, along with the eastern part of the Gourma. The rocks of this region are generally better exposed than in the south and they have been the subject of intensive research over many years.

Figure 6.3 provides a correlation between the northern and southern parts of the eastern Pan African domain in West Africa. The northern part is sometimes called the **Tuareg shield**. The most obvious feature of Figure 6.3 is that the Tuareg shield can be divided into three main blocks, separated and cut by major faults and shear zones. Only the central Hoggar–Aïr segment can properly be correlated with the southern Pan African domain, which it resembles in several respects. Also called the **Suggarian belt**, it is a polycyclic terrane, characterised by elongate low-grade metasedimentary belts of probable Upper Proterozoic age in an older (at least Eburnian) gneiss–migmatite basement, the whole reactivated and intruded by granites around 650–600 Ma ago. Kibaran ages (*c.* 1100 Ma) have also been recorded, though (as in Nigeria) their significance is not clear. There are mainly acid metavolcanics among the metasediments, which are considered to have accumulated in ensialic troughs, the boundaries of which are in part represented by major N–S shear zones.

To the west lies the **Pharusian belt**, comprising two N–S zones of late Proterozoic rocks deformed and metamorphosed in the Pan African. They are separated by a narrow elongate block of cratonic crust, which has granulite facies rocks in it and may be as old as Archaean (*c.* 3000 Ma). This is the **In Ouzzal block**, now bounded by faults and shear zones, and believed to be a thrust block of lower crust. In the western Pharusian zone there are remnants of an older platform succession including quartzites and stromatolitic limestones, which are comparable to those found in the Taoudeni and Volta Basins (Part II). However, the main Pharusian succession comprises volcanic and volcanosedimentary rocks that include andesites and dacites, greywackes, turbidites, conglomerates and pelitic facies. These rocks are all of Upper Proterozoic age and were deformed in the Pan African (*c.* 620 Ma). They were metamorphosed to greenschist and amphibolite grades, with the development of giant Alpine-type nappes and thrust sheets, the overall direction of transport being westward, towards the craton. There are ultramafic masses along the cratonic boundary, which is also associated with positive gravity anomalies. Considerable syntectonic

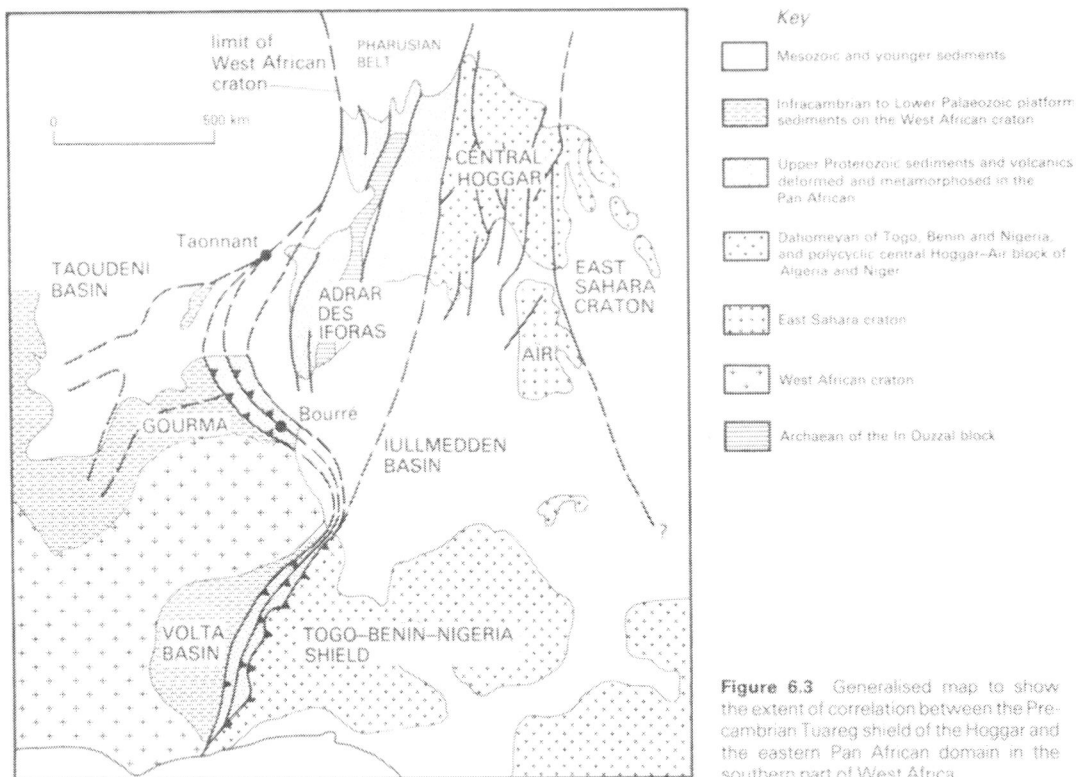

Figure 6.3 Generalised map to show the extent of correlation between the Precambrian Tuareg shield of the Hoggar and the eastern Pan African domain in the southern part of West Africa.

to late-tectonic granitic plutonism occurred in the Pharusian belt (650–530 Ma) and some major N–S shear zones developed. The '**série pourprée**' of the Pharusian belt is regarded as a molassic deposit occupying late-stage graben, and is at least partly of Lower Cambrian age (the 'greenstones' of the Pharusian belt are sometimes called the '**série verte**').

South-west of the main Pharusian belt is the Gourma region, in the east of which is another zone of Pan African deformation and metamorphism. This is a complex region, for the great embayment into the West African craton appears to be related to the development of a large rifted basin extending well into the craton. The overall WSW–ENE trend of the Gourma Trough is marked by positive gravity anomalies which provide some of the evidence for the two branches shown in Figure 6.3. The shape of the basin is further defined by the distribution of sediments within it. The total thickness of those sediments is up to 8000 m, thinning southwestward, where the trough narrows and shallows into and on

to the craton. Conglomerates and sandstones at the base area are overlain by silts and shales, and then by carbonates, including stromatolitic facies and resembling those of the lower parts of the Taoudeni Basin succession (Part II). A phase of clastic sedimentation followed, prograding towards the north-east and filling the trough. The sediments at the deeper northeastern end of the graben were powerfully affected by the Pan African orogeny, with the formation of high-grade metamorphic rocks and southwesterly directed nappes and thrust sheets. An Eburnian inlier at Bourré is interpreted as a horst block, a large cratonic slice thrust upwards from deeper levels during the deformation. It consists of amphibolite facies Birimian metasediments intruded by an Eburnian granite (c. 2080 Ma). Ultramafic bodies which may be ophiolite fragments are encountered in this region, as they are further north. Deformation and metamorphism are much diminished south-east of the thrust zone, that is, where the sediments rest upon cratonic basement, and here they are overlain unconformably by late Infracambrian sandstones. These are not a molasse deposit, however, for they are part of the main Taoudeni Basin succession and show evidence of northeastward progradation (Part II).

East of the central Hoggar–Aïr block is an apparently older terrane, with largely undeformed and unmetamorphosed platform sediments and molasse deposits, the whole intruded by granites. Relatively little appears to be known about this region, but it is identified as part of an **East Saharan craton**. It may be substantially younger than the West African craton, however, for sparse age data suggest that it was not stabilised until about 730 Ma ago, only a short time before the main phase of Pan African activity.

6.4 Pan African plate tectonics east of the craton

The Pan African event in the Tuareg shield (Fig. 6.3) is seen by the majority of those who have worked there as a fully developed continental collision orogeny of the kind illustrated in Figure 2.2, involving the West African and East Saharan cratons, and on a scale comparable with that of the Himalayan mountain chain. The Pharusian belts are believed to represent the site of a late Proterozoic ocean basin, in which deep-water sediments accumulated off the margin of the West African craton, and in which volcanic island arcs developed. The In Ouzzal block was presumably originally one or more microcontinental fragments in this ocean. The Gourma Basin is interpreted as an aulacogen, an intracontinental rift extending inland from the continental margin, i.e. southwestward from the margin of the craton. The time of its initiation is placed at 850–800 Ma ago, as

the late Proterozoic ocean basin east of the craton.

The evidence for a major continental collision event in the Pan African is less overwhelming in the Dahomeyan domain. Nonetheless, the whole of the eastern craton boundary is marked by a zone of positive gravity anomalies (cf. Sec. 5.2), and ophiolite fragments have been identified at intervals along it. In addition, the regional fracture pattern (Sec. 6.3.4) is consistent with a plate collision model for the whole of this eastern Pan African domain, and it seems logical to identify the Togo belt as a southern extension of the Pharusian belt.

Figure 6.4 shows the main zone of supracrustals to be ensialic rifted troughs. However, it is also possible that the Togo–Benin–Nigeria swell, and perhaps the central Hoggar–Aïr block as well, represents aggregates of continental fragments, along with island arcs and inter-arc basins, some ensialic, others ensimatic, swept together by subduction and collision to form a coherent continental block accreted on to the West African craton margin. Such a model could perhaps account for a pre-Pan African (Kibaran) episode of deformation, metamorphism and reactivation.

The relative merits of different interpretations of the Pan African event in West Africa can be endlessly debated, but do not justify further discussion here. Figure 6.4 must be seen as just one possible model.

6.5 Economic potential of the eastern Pan African domain

The Pan African terranes in West Africa contain fewer large mineral deposits than the older Birimian and Archaean rocks of the craton. There seem to be two main factors responsible for this. First, the

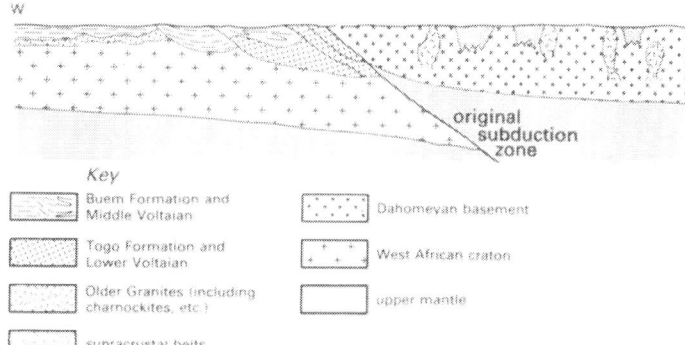

Figure 6.4 Schematic cross section (not to scale) to illustrate a possible plate tectonic interpretation for the southern part of the eastern Pan African domain in West Africa.

Key

Buem Formation and Middle Voltaian

Togo Formation and Lower Voltaian

Older Granites (including charnockites, etc.)

supracrustal belts

Dahomeyan basement

West African craton

upper mantle

proportion of mafic to ultramafic rocks among the supracrustals is relatively small. Secondly, the polycyclic nature of the basement has resulted in large tracts becoming barren of useful mineralisation, although the Pan African reactivation did cause enrichment of some elements in a few places. In general, there is a considerable variety and abundance of relatively minor mineral deposits that offer scope for the operation of small-scale mining enterprises, producing minerals, both to fill local needs and for export. As the mineral deposits are generally small and some occur over wide areas, the distribution of most of them can be described with reference to Fig. 6.1 and do not justify the inclusion of a separate map.

Gold was extracted for many years from several parts of the supracrustal belts in the western half of Nigeria. Recorded total production from the Nigerian goldfields is about 300 000 oz (*c.* 9000 kg) since the early 1920s, but the real total must be much more than this. At the height of mining activity in the early 1930s, annual production was of the order of 30 000–40 000 oz, which is less than a tenth of the annual production from the Ghana goldfields (Sec. 4.7.1). Gold mining in Nigeria declined during and after World War II and remains at a low level.

Most of the gold production was from alluvial and eluvial placers in river systems draining supracrustal belts in three main areas: around Maru and Anka in the north-west, extending southward to beyond Zuru; in the belts west of Kaduna; and in the south-west in the Ife and Ilesha region. The primary gold is in the form of veins, stringers, lenses, reefs and similar bodies of quartz, quartz–feldspar and quartz–tourmaline rock in both supracrustals and basement. The veins range in thickness from several centimetres to a few metres, and in length up to several hundred metres, often displaying lenticular or pinch-and-swell (boudinage) structure. They are invariably steeply inclined and may occur as isolated bodies or as parallel or *en echelon* vein systems. The veins are commonly associated with fractures and shear zones, and are mainly concordant with regional foliation trends, but often also cross cutting.

There is no systematic correlation between gold mineralisation and the type of country rock, no obvious association of gold-bearing veins with Older Granites or particular lithologies among the supracrustal rocks. However, it is noteworthy that the main areas of gold occur near major faults and shear zones (Sec. 6.3.4). In addition, a very crude spatial zonation has been recognised: in the south the veins tend to be of pegmatitic type (notably the Iperindo vein near Ilesha), further north are quartz veins carrying sulphides along with the gold (mainly galena, but also variable amounts of pyrite, arsenopyrite, and possibly chalcopyrite), and further north still, simple gold–quartz veins predominate.

The primary gold veins are all small and were mostly worked at the surface, only a few being large enough to justify underground mining. Only relatively few of these primary gold deposits were found, however, and it is likely that a high proportion of the placer gold has come from small disseminated pods and veinlets with low gold contents that became concentrated by alluvial processes.

Gold is also known from a few places in the Togo belt, associated with the thrust zones. The largest deposit is at Perma in northern Benin, where there has been some mining of alluvial and eluvial concentrations. These were last mined in 1956, but recoverable reserves are estimated at 3000 kg, and production may start again in the near future.

A broad belt of *tin–tantalum–niobium*-bearing pegmatites extends northeastwards for some 400 km from near Ife to the Wamba–Jemaa area just southwest of the Jos Plateau. They are emplaced more or less conformably into a variety of rock types, including Younger Metasedimentary mica–schists, quartzites, amphibolites and calc-silicate rocks, as well as basement gneisses and migmatites with their Older Metasediment relics.

Many of the mineralised pegmatites are massive bodies that have pronounced pinch-and-swell (boudinage) structures, the swellings being loci of intense albitisation of the feldspars and commonly of rich mineralisation. Most pegmatites in the Nigerian basement are dominated by quartz and microcline, commonly accompanied by varying proportions of oligoclase, biotite, muscovite and tourmaline, and they are barren of economic minerals. In general, it is only where late-stage Na-rich hydrothermal solutions have produced *either* secondary albitisations of the feldspars *or* quartz–mica **greisens**, that mineralisation can be expected. In addition to the main economic minerals, cassiterite, columbite and tantalite, there is a host of accessories, including *scheelite, beryl, apatite, monazite*, the Li-rich mica *lepidolite*, black and pink-green *tourmaline* and gem quality blue *gahnite* (Zn-rich spinel).

Field relations indicate that emplacement of the pegmatites was part of the Pan African Older Granite cycle, but it must be emphasised that there is commonly no obvious spatial relationship with granitic intrusions. This is borne out by age determinations. Rb/Sr measurements on pegmatites give ages of around 550 Ma, but nearby granites can be as old as 650 Ma, that is, up to 100 Ma older. Isotopic data indicate a considerable crustal involvement in the generation of these rocks. K/Ar measurements on the micas of some pegmatites have given ages as young as Mesozoic in some cases. This has been ascribed to a reheating episode during emplacement of the Younger Granites (Ch. 15).

It is noteworthy that the pegmatite belt cuts at a high angle across the regional grain of the Pan African rocks, even though individual bodies are more or less conformable with it. The reasons for this distribution pattern are still not clear, but it does appear to be real and not an artefact of geological mapping. Mineralised pegmatites have been found outside the belt, but they are small and not especially rich. It may be merely a coincidence that the trend of the belt is parallel to that of the Benue Trough (Sec. 11.1).

Most of the ores are taken from eluvial and alluvial placers, but the pegmatites themselves are also worked. The pegmatites provide nearly all of the tantalum production in Nigeria (never more than about 20 tonnes annually, and in general averaging about 5 tonnes per year), but only a small proportion of the tin and niobium, the remainder being produced from the Younger Granites of the Jos Plateau (Part III).

The possibility that vein-type *uranium* mineralisation is associated with basement rocks and Older Granites has stimulated exploration from time to time, but so far with little success. The discovery of such deposits in the Poli area of northern Cameroun, east of the Mambilla Plateau, however, suggests that others may be found; uranium showings have also been reported from the Tessalit region of the western Adrar des Iforas in Mali, and from some localities in Nigeria.

Iron ores are the only other metallic minerals that can be presently considered of economic interest in this age province. In some supracrustal belts (e.g. the Maru belt) there are banded quartz–haematite rocks, sometimes magnetite-bearing, associated with garnet–grunerite–schists, amphibolites, phyllites and quartzites. They resemble the banded iron formations (itabirites) of the Archaean terrane (Sec. 3.9.2), but are much smaller and leaner, with average grades not exceeding 40% Fe at best. A certain amount of supergene secondary enrichment has occurred in the surface weathering zone, but this is not deep, and the deposits cannot be regarded as a promising economic prospect.

Similar considerations may apply to deposits of banded iron formation type in rocks of both the Buem and Togo Formations in the northern part of the Togo belt. A deposit estimated to contain some 95 million tonnes of iron ore of unspecified grade has been reported near Bandjeli in the west. Further east, in the region round Dako, magnetite–haematite(–limonite) ores grading 40–45% Fe are associated with quartzites and mica-schists, some of which contain chlorite, garnet and epidote. There are several separate deposits each containing some hundreds of thousands of tonnes of ore.

However, there are richer and purer itabirite-type ores interbedded among basement gneisses and forming prominent ridges near Okene (south-east of Kabba) in southern Nigeria. These are probably Older Metasediment relics and they could be as old as Archaean (and perhaps can be correlated with the iron ores of Liberia and Guinea; Sec. 3.9.2). There are four such ridges, of which Itakpe Hill, some 15 km north-east of Okene, has been investigated in most detail. It has the form of an isoclinal fold oriented WNW–ESE and closing to the south-east, with an overall southerly dip. The ores are massive, banded (gneissic or schistose), depending partly on composition and partly on the degree of recrystallisation that the rocks have experienced. The main ore minerals are magnetite and haematite. At least 150 million tonnes of 35–50% Fe have so far been proved. The rocks are very quartz-rich in some places, but undesirable impurities such as S and P are absent.

These ores will provide raw materials for the Nigerian iron and steel industry, which started production early in 1982, at Aladja, using natural gas as fuel. There is also a major steel complex at Ajaokuta (on the Niger River) and others are being built. Two rolling mills were commissioned in 1983. Initial planned capacity is of the order of 1 million tonnes per year, rising eventually to around 5 million, using ores from Liberia and Brazil as well as from Okene. Output in 1983 totalled over 700 000 tonnes of iron and steel products.

Magnetite–quartzites in the Kribi region of southern Cameroun resemble the Okene ores. Grades are around 40% Fe, though secondary haematite enrichment has raised this to 70% in places. Deposits of similar type, though smaller, are believed to occur in other areas of the Pan African basement.

Manganese does not occur in any quantity in the

Pan African domain. Ten million tonnes of ore have been reported near Bayega in Togo, and in Nigeria some very small lenses of gondite-type ore have been found, with limited supergene enrichment. There must be other deposits but probably nothing on the scale encountered in Birimian rocks (Sec. 4.7.2). *Bauxite* is another mineral that is so far not known from the Pan African terrane, except for a small deposit at Mt Agou in Togo, where it appears to be associated with a charnockitic complex. *Rutile* is found as alluvial placers in and near the Togo belt and occasional small concentrations have formed in other basement localities, probably derived from otherwise barren pegmatites.

So far as other ferrous and base metals are concerned, the overall prospects in the Pan African domain are generally similar to those in the older rocks of the craton. There are many small and scattered concentrations, but none of significant size has so far been discovered.

Small *chromite* deposits occur among the ultramafic masses associated with the Togo belt and at the adjacent Dahomeyan basement, brought up from deeper levels by thrusting. In Nigeria minor concentrations of chromite occur in serpentinites in supracrustal belts and emplaced along shear zones in the basement. Such bodies might be more favourable as sources of *asbestos*, *talc* and *magnesite*, occurrences of which have been found in a few places, though so far only in small amounts and generally of poor quality. *Nickel* anomalies are associated with these ultramafic masses, but the prospects for significant concentrations of this metal are not promising.

An occurrence of *copper* and *vanadium* is recorded at the northern end of the Togo belt, associated with basic rocks, and copper showings are reported in south-west Benin. Otherwise, these metals, along with *lead* and *zinc*, are virtually absent from the Pan African terrane. Small amounts of chalcopyrite occur in some amphibolites and altered gabbros and there is galena in some gold-bearing veins, but all these showings are small and disseminated.

In addition to asbestos, talc and magnesite that have already been mentioned, industrial minerals occur in various places. *Kyanite* is known from at least one locality, and there may also be concentrations of *sillimanite*, especially among the higher-grade aluminous schists of the southern belts in Nigeria. *Mica* is extracted from some of the pegmatites and coarser micaceous schists.

In a few places, granitic rocks have been altered by hydrothermal fluids forming deposits of almost pure white *kaolinite*, the largest of which so far found is at Kankara and has been extensively worked.

There are numerous occurrences of *marble*, mainly Older Metasediment relics in the basement and mostly in southern parts of Nigeria, but also in Togo and Benin. The deposit at Jakura is exploited partly for the construction industry, partly to provide fluxing material for the iron and steel complexes. Some of the marbles are fairly pure and calcitic, but others are dolomitic and unsuitable for cement manufacture.

As elsewhere, an abundance of crushed rock and aggregate for construction purposes is available throughout the basement areas, particularly in places where the superficial weathering and alluvial deposits are thin and bedrock comes to the surface, notably around inselbergs.

Bibliography

Affaton, P., J. Sougy and R. Trompette 1980. The tectono-stratigraphic relationships between the Upper Precambrian and Lower Palaeozoic Volta Basin and the Pan-African Dahomeyide orogenic belt (West Africa). *Am. J. Sci.* **280**, 224–48.

Ajibade, A. C. 1976. Provisional classification and correlation of the schist belts in north-western Nigeria. In *Geology of Nigeria*, C. A. Kogbe (ed.). Lagos: Elizabethan Press.

Ajibade, A. C., W. R. Fitches and J. B. Wright 1979. The Zungeru mylonites, Nigeria: recognition of a major tectonic unit. *Rev. Géol. Dyn. Géog. Phys.* **21**, 359–63.

Ball, E. 1980. An example of very consistent brittle deformation over a wide intracontinental area: the late Pan-African fracture systems of the Tuareg and Nigerian shields. *Tectonophysics* **61**, 363–79.

Batchelor, R. A. and J. A. Kinnaird 1984. Gahnite composition compared. *Mineral. Mag.* **48**, 425–9.

Bertrand, J. M. and R. Caby 1978. Geodynamic evolution of the Pan-African orogenic belt: a new interpretation of the Hoggar shield (Algerian Sahara). *Geol. Rdsch.* **67**, 357–88.

Black, R. 1980. Precambrian of West Africa. *Episodes* **1980**, no. 4 (Dec.), 3–8.

Black, R., R. Caby, A. Moussine-Pouchkine, R. Bayer, J. M. Bertrand, A. M. Boullier, J. Fabre and A. Lesquer 1979. Evidence for late Precambrian plate tectonics in West Africa. *Nature* **278**, 223–6.

Blanchot, A., J. P. Dumas and A. Papon 1972. *Carte géologique de la partie méridionale de l'Afrique de l'Ouest.* Paris: BRGM.

Caen-Vachette, M., K. J. M. Pinto and M. Roques 1979. Plutons éburnéens et métamorphisme Pan-Africain dans le soele cristallin au Togo et au Bénin. *Rev. Géol. Dyn. Géogr. Phys.* **21**(5), 351–8.

Cahen, L. and N. J. Snelling 1984. *Geochronology and evolution of Africa.* Oxford: Oxford University Press.

Dessauvagie, T. H. J. 1974. Geological map of Nigeria, 1 : 1 million. *Nigerian J. Mining Geol.* **9** (1 & 2).

Eborall, M. I. 1976. Intermediate rocks from Older Granite complexes of the Bauchi area, northern Nigeria. In *Geology of Nigeria*, C. A. Kogbe (ed.). Lagos: Elizabethan Press.

Elueze, A. A. 1982. Geochemistry of the Ilesha granite gneiss in the basement complex of southwestern Nigeria. *Precambrian Res.* **19**, 167–77.

Fabre, J. 1976. *Introduction à la géologie du Sahara Algérien.* Algiers: SNED.

Fabre, J. 1982. Pan-African volcano-sedimentary formations in the Adrar des Iforas (Mali). *Precambrian Res.* **19**, 201–14.

Falconer, J. D. 1911. *The geology and geography of northern Nigeria.* London: Macmillan.

Grant, N. K. 1969. The late Precambrian to early Palaeozoic Pan-African orogeny in Ghana, Togo, Dahomey, and Nigeria. *Geol. Soc. Am. Bull.* **80**, 45–56.

Grant, N. K. 1970. Geochronology of Precambrian basement rocks from Ibadan, southwestern Nigeria. *Earth Planet. Sci. Lett.* **10**, 26–38.

Grant, N. K. 1973. Orogeny and reactivation to the west and southeast of the West African craton. In *The ocean basins and margins*, vol. 1, A. E. M. Nairn and F. G. Stehli (eds). New York: Plenum.

Grant, N. K., M. Hickman, F. R. Burkholder and J. L. Powell 1972. Kibaran metamorphic belt in Pan-African domain of West Africa? *Nature (Phys. Sci.)* **238**, 90–1.

Holm, R. F. 1973. Ultramafic and mafic protoclastic intrusions in the Dahomeyan gneiss of Ghana. *Geol. Mag.* **110**, 557–64.

Hubbard, F. H. 1975. Precambrian crustal development in western Nigeria: indications from the Iwo region. *Geol. Soc. Am. Bull.* **86**, 548–54.

Hussein, H. A. 1980. Origin of uranium mineralisation at Goble, Foli district, north Cameroun. *Ann. Geol. Surv. Egypt* **10**, 863–9.

Jackson, B. 1982. Gem quality gahnite from Nigeria. *J. Gemmol.* **18**, 265–76.

Kpalma, A. S. and F. K. Seddoh 1983. Les minéralisations de type itabirite dans l'Atacorien du Togo. *J. Afr. Earth Sci.* **1**, 199–212.

Lasserre, M. and D. Doba 1979. Migmatisation d'âge panafricain au sein des formations camerounaises appartenant à la zone mobile de l'Afrique centrale. *C.R. Som. Soc. Geol. Fr.* **2**, 64–8.

McCurry, P. 1976. The geology of the Precambrian to Lower Palaeozoic rocks of northern Nigeria – a review. In *Geology of Nigeria*, C. A. Kogbe (ed.). Lagos: Elizabethan Press.

Matheis, G. and M. Caen-Vachette 1983. Rb–Sr isotopic study of rare-metal and barren pegmatites in the Pan-African reactivation zone of Nigeria. *J. Afr. Earth Sci.* **1**, 35–40.

Mücke, A. and C. Okajeni 1984. Geological and ore microscopic evidence on the epigenetic origin of the manganese occurrences in northern Nigeria. *J. Afr. Earth Sci.* **2**, 204–26.

Odeyemi, J. B. 1981. A review of the orogenic events in the Precambrian basement of Nigeria, West Africa. *Geol. Rdsch.* **70**, 879–909.

Okeke, P. O. and B. E. Bafor 1984. A rare-earth element geochemical study of some Precambrian rocks from the metasedimentary belt of northwestern Nigeria. *J. Afr. Earth Sci.* **2**, 229–34.

Olade, M. A. 1976. Metallic mineral resources and exploration potential of Nigeria. In *African geology* (Khartoum, 1976), Tsegaye Hailu and A. Badejoko (eds). Ibadan: Geol. Soc. Africa.

Olade, M. A. 1978. General features of a Precambrian iron ore deposit and its environment at Itakpe ridge, Okene, Nigeria. *Trans. IMM* **B87**, 1–9.

Olade, M. A. 1980. Precambrian metallogeny in West Africa. *Geol. Rdsch.* **69**, 411–28.

Olade, M. A. and A. A. Elueze 1979. Petrochemistry of the Ilesha amphibolites and Precambrian crustal evolution in the Pan-African domain of SW Nigeria. *Precambrian Res.* **6**, 303–18.

Oyawoye, M. O. 1972. The basement complex of Nigeria. In *African geology*, T. J. J. Dessauvagie and A. J. Whiteman (eds). Ibadan: Ibadan University Press.

Rahaman, M. A. 1976. Review of the basement geology of south-western Nigeria. In *Geology of Nigeria*, C. A. Kogbe (ed.). Lagos: Elizabethan Press.

Rahaman, M. A., W. O. Emofurieta and M. Caen-Vachette 1983. The potassic-granites of the Igbeti area: further evidence of the polycyclic evolution of the Pan-African belt in southwestern Nigeria. *Prec. Research* **22**, 75–92.

Trompette, R. 1979. Les Dahomeyides au Bénin, Togo et Ghana: une chaîne de collision d'âge pan-africain. *Rev. Géol. Dyn. Géog. Phys.* **21**, 339–49.

Turner, D. C. 1983. Upper Proterozoic schist belts in the Nigerian sector of the Pan African province of West Africa. *Prec. Research* **21**, 55–79.

Vachette, M. 1975. Age Pan-Africain des granites de Sinendé, Savé et Fita (Dahomey). *C.R. Acad. Sci. Paris* **281**, 1793–5.

van Breemen, O., R. T. Pidgeon and P. Bowden 1977. Age and isotopic studies of some Pan-African granites from north-central Nigeria. *Precambrian Res.* **4**, 307–19.

7 The Precambrian of West Africa – synthesis and review

SUMMARY

The similarities between the three major age provinces in the West African Precambrian are more striking than the differences, though the rocks span more than 2000 Ma of geological time. As there is good evidence that plate tectonics played a major part in the evolution of the Pan African domains, the resemblance to older terranes suggests that some form of plate tectonics has been active since the Archaean.

The distribution of mineral deposits, however, has had more to do with levels of erosion, relative abundance of different rock types, climatic changes, erosion levels and intensity of weathering than with the fundamentals of global tectonics. Rock associations remain the most important control on the occurrence of mineral deposits.

7.1 The geological picture

It is worth reiterating a number of generalisations that appear to be valid for the whole of the Precambrian terrane in West Africa, irrespective of age. They apply to rocks that span some 2000 Ma.

(a) There is a tripartite division into basement, supracrustals and granitic and related intrusions.

(b) Overall structural trends fluctuate about a generally NNE–SSW direction. Structures are commonly upright, with steep dips and tight to isoclinal folding. Flat-lying structures are found in places, but they do not appear to be common.

(c) The relative proportions of areas underlain by supracrustals and basement are smaller in Archaean and Pan African domains than in the Lower Proterozoic (Birimian) domain. The proportion of different supracrustal lithologies varies both within and between the different domains: volcanics predominate in the western Archaean and eastern Birimian provinces, whereas sediments prevail elsewhere.

(d) Metamorphic grades are generally similar throughout the Precambrian, with amphibolite to granulite facies rocks characterising the basement, greenschist to amphibolite facies more typical of the supracrustals. There are often steep metamorphic gradients between basement gneisses and migmatites and supracrustal belts of schists and phyllites.

(e) There is no conclusive evidence of any significant difference in the volumes of syntectonic to late-tectonic intrusions (mainly granitic) emplaced during each major thermotectonic event, between one age province and another. Nor are there obvious signs of compositional trends in space or time, either within or between the different age provinces.

(f) Major fractures and shear zones, many with an overall NNE–SSW trend, appear to be a feature of the whole region. They may represent deep-seated lines of weakness, repeatedly rejuvenated during successive thermotectonic events.

As the rocks have similar metamorphic grades, they formed under comparable conditions of temperature and pressure, which in turn implies comparable depths of burial. The comparison can only be approximate, for Archaean thermal gradients were steeper than those of the late Proterozoic, because there were more radioactive heat-producing elements in the crust. Comparable metamorphic grades were presumably reached at shallower depths in the Archaean than in the late Proterozoic. Nonetheless, the original depth of burial of the Precambrian rocks was probably between about 10 and 30 km, and this represents the amount of erosion that must have occurred since their formation. Yet at the present time, Archaean rocks more than 2500 Ma old are exposed at about the same level of erosion as Pan African rocks about 500 Ma old. The implication is that erosion rates must have been very high during and immediately after each thermotectonic event or orogeny, and declined greatly thereafter. This is supported by evidence of very high initial rates of erosion in the present-day Alps and Himalayas.

7.2 The tectonic patterns

If similar geological relationships result from similar causal mechanisms, it seems likely that some form of plate tectonics was already in operation during Archaean times. The Pan African mobile belt is similar in age to early Phanerozoic orogenic belts in other parts of the world, for which generally accepted plate tectonic models have been proposed. If the Pan African thermotectonic event was the result of a 'plate tectonic orogeny', then the older terranes were presumably formed by similar processes. Even so, the precise nature of those processes remains a field ripe for speculation. For example:

(a) The rates of movement of Archaean lithospheric plates must have been substantially greater than at present, and the sizes of those plates probably smaller, because the Archaean Earth was a good deal hotter than the present-day one. However, at least in the later part of the Archaean, the thickness of continental crust and the styles of deformation do not appear to have been greatly different from what they are now – for instance, horizontal thrusts and fold movements have been recognised in some Archaean terranes (other than the Kasila belt of Sierra Leone).

(b) It is possible to recognise three main kinds of orogenic belt in Phanerozoic terranes (Fig. 7.1). There are gradations between these types, and features of each can be discerned in different parts of the West African craton and the adjacent Pan African belts.

(c) It has also been proposed that in other parts of the African continent, Pan African terranes represent an aggregation of island arcs and continental fragments swept together by subduction and accretion into a single continental block. If that were to prove valid for the eastern domain of the Pan African in West Africa, then by implication it could also hold for the older rocks of the craton, because of their overall similarities to this Pan African terrane.

While palaeomagnetic data do not appear to permit the opening and closure of large ocean basins during the Precambrian, they do not preclude the development of small ensimatic

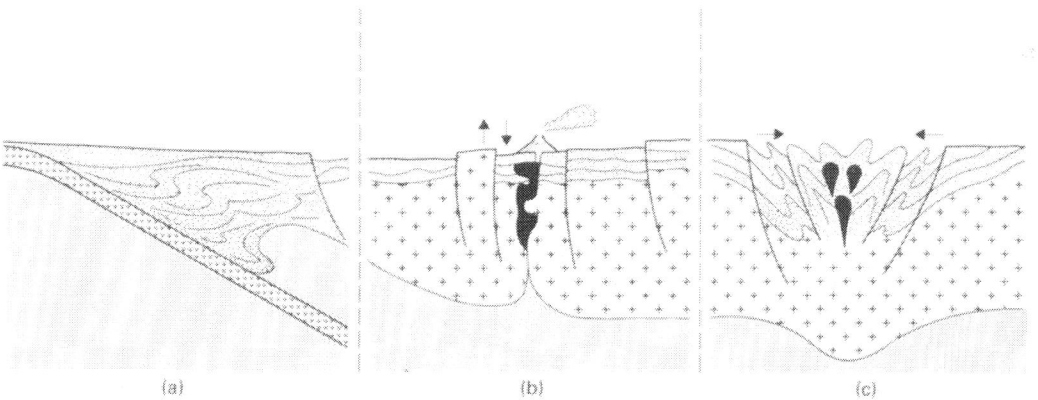

Figure 7.1 Diagrammatic representations of the three main types of orogenic belt recognised in Phanerozoic terranes. (a) *Alpinotype:* Sedimentation is in a marginal trough. Orogeny involves mainly underthrusting and crustal shortening, with development of low-temperature/high-pressure metamorphism, emplacement of ophiolites, and intrusion of gabbros and ultramafic rocks rather than granites. Named for the Alps of southern Europe. (b) *Andinotype:* Sedimentation is in fault-bounded troughs near the continental margin. Orogeny involves mainly vertical movements with little crustal shortening, low-temperature/high-pressure metamorphism, eruption of andesites (which may also accompany sedimentation), and intrusion of granites and related rocks. Named for the South American Andes. (c) *Hercynotype:* Sedimentation is in intracontinental basins. Orogeny involves both vertical and horizontal movements with some crustal shortening, moderate- to high-pressure/temperature metamorphism, with formation of migmatites and intrusion of granites and related rocks. Named for the Hercynian orogenic belt of northern and central Europe.

71

troughs and back-arc basins. They are also consistent with oblique rather than 'head-on' collisions between crustal blocks and fragments and island arcs. Major shear zones would form in this way, but it must be borne in mind that they can also develop as a result of direct collision, as seen in the present-day Himalayan mountain belt.

(d) An additional variable is the rate and angle of subduction beneath island arcs and continental margins, which can affect patterns of metamorphic rejuvenation and related magmatism in the overlying crust.

(e) Finally there remains the question of the rate of growth of continental crust through geological time. The evidence of repeated reactivations and generation of granitic magmas by crustal remelting that seems to characterise the Precambrian of West Africa supports the proposition that most of the world's continental crust was in existence by the end of the Archaean. It does not constitute proof, however, and there is still much debate about the mechanisms of crustal growth and the relative importance of juvenile contributions from the mantle versus recycling of crustal material.

7.3 The mineral deposits

Contrasted and conflicting ideas about the possible role of plate tectonics during the Precambrian remain of academic interest in relation to the distribution of mineral deposits in West Africa. The most important mineral deposits in the Precambrian of West Africa as a whole are iron and manganese ores, gold and diamonds. The first two owe their origin more to accidents of marine geochemistry than to tectonic forces, the manganese ores requiring in addition special weathering conditions to make them exploitable. Gold is an elusive element prone to redistribution by hydrothermal processes and subsequent concentration into alluvial placers, and its ultimate origin is not easy to determine. Diamonds come from kimberlites that originate at great depths, but the diamonds in the Birimian are found in sediments and their origin cannot yet be related to any primary source region. Bauxite is another ore whose occurrence and distribution is due mainly to favourable climatic conditions acting on a variety of rock types. Plate tectonic models have been used to *explain* the distribution of different kinds of mineral deposits, particularly of base metals such as copper, lead–zinc and tin, but few predictive models exist.

Rock associations are still the best guide to prospecting. Thus, tin is not found in ultramafic rocks, chromium is not found in granites. Similarly, coal and petroleum and phosphates are not sought in igneous and metamorphic terranes. They are to be found among unmetamorphosed and largely undeformed sediments, such as those that occupy the great sedimentary basins of West Africa.

Bibliography

Bickle, M. J., L. F. Bettenay, C. A. Boulter, D. I. Groves and P. Morant 1980. Horizontal tectonic interaction of an Archaean gneiss belt and greenstones, Pilbara block, Western Australia. *Geology* **8**, 525–9.

Cobbing, E. J. 1978. The Andean geosyncline in Peru and its distinction from Alpine geosynclines. *J. Geol. Soc. Lond.* **135**, 207–18.

Condie, K. C. 1980. Origin and early development of the Earth's crust. *Precambrian Res.* **11**, 183–97.

Condie, K. C. 1982. Early and middle Proterozoic supracrustal successions and their tectonic settings. *Am. J. Sci.* **282**, 341–57.

England, P. 1980. Heat flow and the deep structure of continents, *Nature* **285**, 611–12.

Gass, I. G. 1977. The evolution of the Pan African crystalline basement in NE Africa and Arabia. *J. Geol. Soc. Lond.* **134**, 129–38.

Gass, I. G. 1980. *Evolutionary model for the Pan-African crystalline basement.* Inst. African Geol. Bull., no. 3, pp. 11–20.

Jackson, J. and R. Muir Wood 1980. The Earth flexes its muscles. *New Scientist* **88**, 717–20.

Kröner, A. 1981. Precambrian crustal evolution and continental drift. *Geol. Rdsch.* **70**, 412–28.

Lambert, R. St J. 1980. The thermal history of the Earth in the Archaean. *Precambrian Res.* **11**, 199–213.

McLennan, S. M. and S. R. Taylor 1980. Th and U in sedimentary rocks: crustal evolution and sedimentary recycling. *Nature* **285**, 621–4.

Neary, C. R., I. G. Gass and B. J. Cavanagh 1976. Granitic association of northeastern Sudan. *Geol. Soc. Am. Bull.* **87**, 1501–12.

Onstott, T. C. and R. B. Hargreaves 1981. Proterozoic transcurrent tectonics: palaeomagnetic evidence from Venezuela and Africa. *Nature* **289**, 131–6.

Pitcher, W. S. 1979. The nature, ascent and emplacement of granitic magmas. *J. Geol. Soc. Lond.* **136**, 627–62.

Tarling, D. H. 1980. Lithosphere evolution and changing tectonic regimes. *J. Geol. Soc. Lond.* **137**, 459–68.

van Breemen, O. and B. J. Bluck 1981. Episodic granite plutonism in the Scottish Caledonides. *Nature* **291**, 113–17.

Part II
SEDIMENTARY BASINS IN WEST AFRICA

8 Introduction to sedimentary basins

SUMMARY

An older and a younger group of sedimentary basins can be distinguished in West Africa, and the basins are of two main kinds. Intracontinental basins are relatively broad and shallow with a thin sediment fill, generally not much more than 5 km. Coastal basins are normally rather narrow and deep, with sediments up to 10 km thick or more.

The older group comprises the Taoudeni and Volta Basins, both of the intracontinental type, developed on the West African craton. Lower parts of the sequences in them are the platform sediments that were deposited contemporaneously with the sediments and volcanics of the adjacent mobile belts deformed in the Pan African. Palaeozoic sediments record an overall retreat of shelf seas from the craton, following the Pan African event.

The younger basins developed mainly in the Mesozoic.

The large and subcircular Iullmedden and Chad Basins are intracontinental, and so are the more linear Bida Basin and Benue Trough, though the latter is a rifted basin and something of a special case. Continental margin basins include the Senegal Basin and Niger Delta, along with the smaller coastal basins. The seas retreated from the intracontinental basins early in the Tertiary, as the Alpine orogeny commenced in southern Europe. There was regression in the coastal basins at the same time, but sedimentation continued in them offshore.

Several fracture and lineament systems have been identified in West Africa, but there is no evidence of major lateral displacements since the Pan African, and certainly not during the Mesozoic or Tertiary.

8.1 Distribution of the basins

The sediments in the basins rest with major unconformity upon eroded and peneplained metamorphic and igneous rocks formed in the various deformation events described in Part I. The term **basement** can therefore now be used in its wider context to include all rocks below the unconformity. There is a strong contrast between the basement and the overlying sediments, which have experienced minimal deformation and metamorphism. The sediments are in part marine, in part terrestrial (fluviatile and lacustrine). They contain a record of repeated transgressions and regressions of the sea over areas of very low relief. That record extends, with many discontinuities, from the Upper Proterozoic (about 1000 Ma ago) up to the Quaternary and Recent, and different parts of it are preserved in separate basins.

In the southern part of West Africa, the basins may be considered in two broad groups, in terms of age and setting (Fig. 1.4). The older group comprises the Volta Basin and the great Taoudeni Basin (of which only the southern half is dealt with here), with the Bové Basin extension in the south-west and the Gourma depression in the east. They contain sediments ranging mainly from Upper Proterozoic to Palaeozoic in age, and include facies interpreted as the products of two ancient glaciations, one in the Infracambrian, the other in the Upper Ordovician.

The younger group includes the large Iullmedden and Chad Basins in the north, with their elongate southward extensions along the Niger valley (the Bida Basin) and the Benue Trough. The Niger Delta and the Senegal Basin, along with the smaller coastal basins (e.g. Dahomey Basin) also belong to this group. Sedimentation in the younger basins commenced in the Mesozoic and continued into the Quaternary.

Both the older and younger intracontinental basins developed by epeirogenic warping or stretching and rifting of tectonically stabilised crust, on wavelength scales of hundreds of kilometres and amplitudes of only a few kilometres. A page-size cross section for the Taoudeni Basin, for example, would show the full thickness of the sediments as little more than 1 mm at true scale.

Subsidence in these great shallow depressions must have continued for tens to hundreds of millions of years. However, the lithology of the sediments is everywhere consistent with water depths not more than a few hundred metres.

The thickness of sediments generally increases inwards from the margin, but it is not necessarily greatest at the centre of the basin. Some parts may have subsided more than others, forming local 'deeps'. Subsidence in the basins was accompanied by differential or relative uplift of the margins. The removal of material by erosion from topographically higher regions round a basin and its accumulation within the basin could be a factor in maintaining the differential vertical movements: the margins tending to rise isostatically as material is removed from them by erosion, and the load of sediments accumulating in the basin helping to depress the floor still further.

Differential vertical movements can also occur within such basins, so that sedimentation may continue in one part, while others are subjected to erosion. However, the overall uplift that terminates sedimentation altogether must take place on a regional scale, with relatively little crustal flexuring or warping, so that the form of the basins is preserved. They are merely raised a few tens or hundreds of metres above sea level. Such a phase of overall regional uplift must have occurred in West Africa at the end of the Cretaceous, for example, to bring marine sedimentation to an end.

8.2 Older basins

The basement to the older basins is cratonic, that is, they developed upon the West African craton. The Upper Proterozoic to lowermost Palaeozoic sediments in the Volta and Taoudeni Basins can be correlated with rocks in the deformed Pan African belts lying to the west and east of the craton. As these are platform sediments deposited upon cratonic crust that was stabilised in the Eburnian event, some 2000 Ma ago, they remained virtually undisturbed by the profound deformation and metamorphism that affected their equivalents in the adjacent mobile belts during the Pan African. However, the uplift that accompanied the Pan African event appears to have caused a regional westward tilting of the craton. After the Pan African, sedimentation all but ceased in the Volta Basin, and marine conditions in the Taoudeni Basin were confined to the western and northern parts. The seas withdrew from the craton after the Hercynian orogeny at the end of the Carboniferous, and thereafter sedimentation in the Taoudeni Basin has been exclusively non-marine in character.

In their general form, the basins are really giant synclines with sediments dipping inwards round their margins. In the large Taoudeni Basin there are some very broad and gentle subsidiary anticlinal and synclinal folds, so that parts of the sequence have been preserved in some areas and eroded away in others. Sediments in the Gourma region are thicker

than elsewhere in the basin, but they are not much deformed where they rest upon cratonic crust.

8.3 Younger basins

After the areas affected by the Pan African event had become worn down and peneplained to about the same level as the craton, fluviatile and lacustrine sediments began to be deposited in a few places. This phase of continental sedimentation continued until the Lower Cretaceous, becoming more extensive and spreading on to the craton in the eastern Taoudeni Basin.

Marine conditions became established only after the break-up of Gondwanaland in the mid-Cretaceous, which was accompanied by one of the most extensive global transgressions in the history of the Earth. Shallow seas flooded large tracts of northern Africa, but the distribution of marine sediments shows that these regions did not include significant areas of the West African craton. Cretaceous marine basins in West Africa were developed almost entirely on Pan African continental crust. They were also relatively short-lived, for, at the beginning of Tertiary times, the Alpine orogeny commenced as northern Africa collided with southern Europe, there was a decrease in the rate of seafloor spreading, and the seas retreated once more. Thereafter, sedimentation was largely of fluviatile and lacustrine type, and it was not confined to the younger basins. The lithologies and often rich faunas of the Upper Cretaceous to early Tertiary marine sediments show that warm and shallow seas were characteristic of this period.

There are two kinds of sedimentary basin in this younger group. The Iullmedden and Chad Basins, as well as the Bida Basin, are broad and shallow synclinal depressions, similar to the older basins. Their present extent is less than it once was, for numerous outliers can be found beyond the main sediment outcrops. The sediments in these basins are virtually horizontal.

The Benue Trough is also an intracontinental basin, but the sediments in it have been strongly folded and faulted, and it was probably a rifted depression, somewhat analogous to the Gourma Trough in the eastern Taoudeni Basin, but with a rather different history, as will become apparent in later sections.

The coastal basins differ from the intracontinental basins in two important ways. Instead of being broad and shallow, they tend to be long and deep, with up to 10 km of sediments in them, sometimes more. They are developed partly on oceanic crust and partly on continental crust, and a glance at Figure 1.4 will show that this continental crust is of different ages, some Pan African, some cratonic. These basins developed mostly as a result of crustal stretching, faulting and subsidence as Africa and South America separated in the mid-Cretaceous, and their mode of origin is thus distinctly different from that of the intracontinental basins. It may be significant that the largest of these coastal basins (the Senegal Basin and Niger Delta) are partly underlain by Pan African continental crust. It is possible that the crust and lithosphere of Pan African terranes was slightly warmer and thinner than that beneath the cratons, because it was the most recently reactivated. It may accordingly have been less brittle and more ductile, so that there was more crustal stretching, with corresponding subsidence over wider areas.

Sedimentation in the coastal basins continues today, but they also record the end-Cretaceous regression, for the upper parts of their successions exposed on land (Fig. 1.4) consist of Tertiary and younger sediments of the coastal estuarine and deltaic environment.

8.4   Crustal lineaments and basin development

Crustal lineaments have been identified in many parts of the world, based on the results of conventional geological mapping. Sometimes the features are obvious enough, but often the evidence for supposed lineament systems is rather weak. The same applies to the lineaments identified in increasing numbers with the help of remote-sensing techniques that allow overviews of large areas of crust to be obtained.

In West Africa, as in other regions, any such lineaments must originate in the basement. Even though mapping of basement terranes in West Africa is still largely of reconnaissance type, enough is known to enable some general observations to be made.

(a) In the basement as a whole from Archaean to Pan African, there appear to be major faults and shear zones that have two main trends:

  (i) Between N–S and NE–SW; displacements are not often observed and

undoubtedly have vertical components, but the lateral movements recorded are typically dextral.

  (ii) Between ENE–WSW and NW–SE, along which observed lateral displacements are typically sinistral.

(b) There is no evidence of significant lateral displacements along any of these basement fractures since the end of the Pan African orogeny, and in many areas probably long before that. On the other hand, there have been numerous vertical displacements along faults originating in the basement, and these have both controlled the development of some basins and subsequently affected sediments laid down in them. It is not known whether this faulting always occurred along older lines of crustal fracture, but it can be presumed that some of it did. The most notable example is probably the **Ngaoundere Fault Zone** in Cameroun, which is a Pan African fault or shear system – apparently with a strong dextral component as well as vertical movements – that was reactivated in the Cretaceous to form elongate fault-bounded sedimentary basins, and again in the Tertiary, leading to deformation of the sediments. It may also have had some influence on the magmatism of the Cameroun line (Part III).

(c) It is possible to draw a pattern of generally NE–SW and NW–SE trending virtually straight lines over a geological map of West Africa, such as Figure 1.4. Such lines can be drawn along the margins of both the larger intracontinental basins and the smaller coastal basins on the one hand, and along the trend of linear basins on the other. Whether such lines have any significance remains a matter of speculation at this stage. It is clear, however, that any lineaments identified in this way have acted *either* as rift faults along and within elongate basins *or* as hinges along the margins of intracontinental or coastal basins. Examples include the margins of the Dahomey Basin and the line of the Benue Trough and its various branches, especially the long line up the Niger River valley, through the Bida Basin and along the western margin of the Iullmedden Basin, including the Gao Trough (Sec. 10.2). No lateral displacements have been observed along these or any other lines, however.

77

(d) The trends of fracture zones in oceanic crust can be projected to intersect continental margins, for they originated as transform faults when continents initially split apart. Displacements along transform faults are mainly lateral, though there are vertical components of movement as well. It cannot be too strongly emphasised that displacements along fracture zones in comparatively very young oceanic crust have only rarely affected continental crust that is hundreds or thousands of millions of years older. Where this has happened, the line or plane of displacement of continental crust is readily apparent as a major fault zone (e.g. the San Andreas Fault or the New Zealand Alpine Fault). Normally the only effects are to be found in the narrow transition zone between continental and oceanic crust, and even here the movements will be mainly vertical and not lateral. For the most part, the only possible correlation between oceanic and continental fracture systems is that ancient lines of weakness in continental crust may determine where the oceanic transform faults develop.

Oceanic crust is nowhere older than about 200 Ma and in the southern Atlantic its maximum age is little more than 100 Ma. Any plate movements that resulted in significant transcontinental shearing along the extension of an oceanic transform fault into West Africa would have had to occur during the last 100 Ma or so. Nothing in the history of the Mesozoic to Tertiary basins of West Africa suggests that this has happened.

Bibliography

Ball, E. 1980. An example of very consistent brittle deformation over a wide intracontinental area: the late Pan-African fracture systems of the Tuareg and Nigerian shields. *Tectonophysics* **61**, 363–79.

Benkhelil, J. 1982. Benue trough and Benue chain. *Geol. Mag.* **119**, 155–68.

Chukwu-Ike, I. M. 1977. A crustal suture and lineament in North Africa. *Tectonophysics* **40**, 375–82.

Furon, R. 1963. *Geology of Africa* (English edn). London: Oliver & Boyd.

McConnell, R. B. 1969. Fundamental fault zones in the Guiana and West African shields in relation to presumed axes of Atlantic spreading. *Geol. Soc. Am. Bull.* **80**, 1775–82.

Neev, D. and J. K. Hall 1982. The Pelusian megashear system across Africa and associated lineament swarms. *J. Geophys. Res.* **87**, 1015–30.

Watts, A. B. 1982. Tectonic subsidence, flexure and global changes of sea level. *Nature* **297**, 469–74.

Williams, C. A. 1981. The evolution of sedimentary basins. *Nature* **292**, 802.

Wright, J. B. 1976. Fracture systems in Nigeria and initiation of fracture zones in the South Atlantic. *Tectonophysics* **34**, T43–47.

9 Infracambrian to Lower Palaeozoic basins

SUMMARY

The Volta and Taoudeni Basins have similar successions, though they are given different names in different places. Sparse radiometric data place the onset of sedimentation at around 1000 Ma ago, and products of a late Precambrian glaciation are recognised in both basins, marked by a distinctive association of tillite, limestone with barite, and chert.

In the Volta Basin, the sediments can be correlated with rocks of the Togo belt, and the main phase of sedimentation was effectively terminated by the Pan African event, though some molassic deposits may be as young as Devonian.

In the Taoudeni Basin, Infracambrian sediments are much thicker in the east than in the west, because the rifted Gourma depression formed much of the eastern part of the basin in the late Precambrian. Uplift following the Pan African event shifted the main area of sedimentation to the western half in the late Palaeozoic, and the sediments were later gently folded by the Mauritanide deformation. There is an Ordovician tillite in the western Taoudeni Basin.

The Bové Basin developed on the cratonic block that separates the Rokelide belt from the south-west trending arm of the Mauritanide belt. There are Palaeozoic sediments in the northern part of the Iullmedden Basin and in some of the coastal basins.

Geological evidence for the Infracambrian glaciation is persuasive but more controversial than that for the Ordovician one. Both are consistent with palaeomagnetic data which place West Africa near the South Pole at the appropriate times.

The main economic minerals in the older basin successions are phosphate and bauxite, but there is some potential for uranium, gold and barite.

9.1 Introduction

Sedimentation in both the Volta and Taoudeni Basins is characterised by deposition on a surface of varied though generally subdued relief, commencing about 1000 Ma ago. Marine conditions persisted until Devonian and Carboniferous times in the west and north of the Taoudeni Basin, but in the east and in the Volta Basin they ended after the uplifts that accompanied the Pan African event. Figure 9.1 and Table 9.1 summarise the main subdivisions and correlations that have been established within and between the Volta and Taoudeni Basins in the southern part of West Africa. Both basins have generally similar successions.

9.2 The Volta Basin

The Volta Basin has a more or less concentric distribution of sediments, because of its overall gently synclinal form. The oldest sediments outcrop round the margins, the youngest occupy a roughly central position.

The Lower Voltaian, the **Dapango–Bombouaka Group**, is dominated by massive cross-bedded feldspathic sandstones. It is responsible for the high ground of the Kwahu Plateau in the south, at the base of which the uneven pre-Voltaian surface can be seen in places. It has been suggested that the basement hills to the south of the plateau may be exhumed pre-Voltaian topography. The group as a whole is correlated with the Togo Formation on the east of the Volta Basin (Sec. 6.2). The Lower Voltaian is virtually flat-lying throughout most of the basin, but becomes relatively intensely folded as the Togo belt is approached.

Sediments of the **Pendjari Group** (Middle Voltaian) generally rest with slight angular unconformity on the Lower Voltaian, and in some places they fill erosional channels that have cut through the older beds, so that they rest directly on the basement. They form much of the low ground dominated by the Oti Plains of northern Ghana and Togo and they are also known as the **Oti Formation** in Ghana.

The basal conglomerate of the Pendjari Group is interpreted as a tillite (in Ghana it may be represented by the **Akroso Conglomerate**). It contains

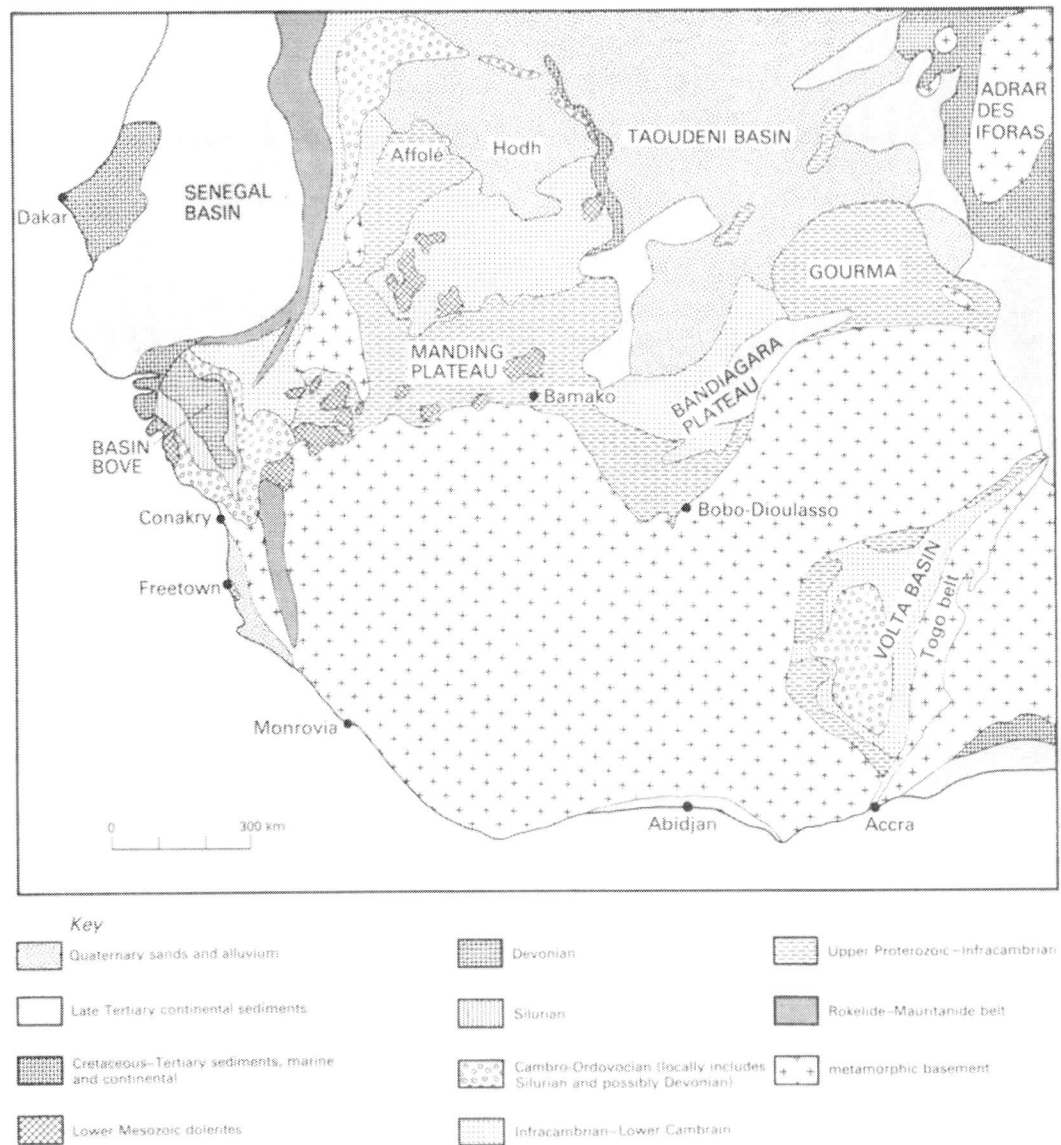

Key

Quaternary sands and alluvium		Devonian		Upper Proterozoic–Infracambrian	
Late Tertiary continental sediments		Silurian		Rokelide–Mauritanide belt	
Cretaceous–Tertiary sediments, marine and continental		Cambro-Ordovocian (locally includes Silurian and possibly Devonian)		metamorphic basement	
Lower Mesozoic dolerites		Infracambrian–Lower Cambrain			

Figure 9.1 Generalised geological map of the Volta Basin and the southern part of the Taoudeni Basin.

boulders up to more than a metre across, subangular to rounded, some with striated, polished or pitted surfaces. It rests on a striated pavement at one locality, where the grooves are oriented 100° and suggest that the sense of movement of the ice was towards the east.

The tillite is succeeded by a variety of sediments including carbonates, often brecciated or slumped, locally barite-bearing and partly **stromatolitic**, along with silexites and silicified argillites. The argillaceous sediments are locally enriched in calcium phosphate which is of considerable economic

80

Table 9.1 Generalised sequences and correlation within and between Taoudeni and Volta Basins.

Age	Supergroup	Taoudeni Basin	Correlation	Volta Basin	Group	Voltaian
Continental intercalaire						
Siluro-Devonian	Supergroup 3	continental sandstones		continental conglomerate and sandstones } molasse	Obosum Group	Upper Voltaian
Cambro-Ordovician c. 450 Ma		marine shales and sandstones (local)		c. 400 m		
		tillite	Upper Ordovician glaciation			
c. 600 Ma		continental sandstones (local)		siltstones, sandstones, greywackes, dolomitic limestones with stromatolites	Pendjari Group	Middle Voltaian
	Supergroup 2	shales, sandstones, dolomitic limestones with stromatolites		silexite (with argillite and phosphate) } the triad		
		silexite }	the triad	limestone and barite }		
		limestone and barite }		tillite }		
		tillite	Vendian glaciation	c. 2000–4000 m		
c. 650 Ma	Supergroup 1	sandstones, shales, stromatolitic and dolomitic limestones		shales, sandstones, limestones (in part stromatolitic)	Dapango-Bombouaka Group	Lower Voltaian
c. 1000 Ma		conglomerates		conglomerates		
				c. 600 m		
Upper Proterozoic and Infracambrian			cratonic basement			
Birimian					Birimian	

importance in places (Sec. 9.6). This association, known as the **triad** by French workers (Table 9.1) is also found in many parts of the Taoudeni Basin and provides an important stratigraphic marker. It is generally not much above 100–200 m thick in total, wherever it is found, and can be appreciably less in places.

The rest of the Pendjari Formation is dominated by shales, siltstones and sandstones, glauconite-bearing in places. The Middle Voltaian is generally correlated with the Buem Formation, on account of the overall lithological similarities between the two groups of rocks.

The Lower Voltaian sediments represent a marine transgression–regression cycle on the craton, while the Middle Voltaian records a glacial event followed by prolonged marine incursion and subsidence of the basin. The thicker Middle Voltaian sediments probably represent a subsiding continental shelf environment. In the eastern parts of its outcrop, adjacent to the Togo belt (Fig. 9.1), the Pendjari Group has been deformed into generally NNE–SSW trending asymmetric folds with southeasterly inclined axial planes.

The **Obosum Formation** (Upper Voltaian) is thickest and coarsest in the south-east. The conglomerates contain pebbles of granite and other igneous rocks, as well as quartzite fragments, and sedimentary structures show the direction of transport to have been from the south-east. There is general agreement that the Obosum beds are a molasse deposit formed by erosion of the Togo belt following its uplift in the Pan African event. Sedimentation may have continued until Devonian times, though there is no direct evidence of this. However, as some of the conglomerates and sandstones have been identified as fluvioglacial in origin, these sediments may also contain a record of the late Ordovician glaciation that is documented in the Taoudeni Basin (Table 9.1).

9.2.1 *Geochronology of the Volta Basin*
Direct evidence of the age of these rocks is confined to a few radiometric age determinations and sparse palaeontological data. However, they support the inferences based on correlations with the Togo belt and the eastern Pan African domain on the one hand, and with the Taoudeni Basin successions on the other. The Middle Voltaian (Pendjari Group) has given ages of 620 Ma (K/Ar on glauconite) and 640 Ma (Rb/Sr on illite), indicating a late Precambrian age, which is consistent with evidence from

stromatolites and microfossils. The age of the tillites is estimated at around 675 ± 25 Ma which places them in the **Vendian**, or uppermost Precambrian, similar to the likely age of the inferred glacigenic sediments at the base of the Rokel River Group (Sec. 5.2).

In summary, the age range of the Lower Voltaian (equivalent to the Togo Formation, Sec. 6.2), is taken to be about 1000–700 Ma, and of the Middle Voltaian (equivalent to the Buem Formation, Sec. 6.2), about 675–570 Ma (cf. Table 9.1). The Upper Voltaian Obosum Formation is a molasse to the Pan African and sediments contributing to it are believed by some workers to range from Ordovician up to Devonian or even younger, a total span of about 450–320 Ma.

9.3 The Taoudeni Basin

The Taoudeni Basin is one of the largest in the world, over 1200 km in diameter. It straddles the whole width of the West African craton, being bordered to the north and south by cratonic 'swells', to the east and west by Pan African belts. The uneven distribution of sediments within it and the gentle folding to which they have been subjected means that there is no simple outcrop pattern as in the Volta Basin. In addition, much of the older succession is obscured by Mesozoic–Tertiary sediments and Quaternary desert sand (Fig. 9.1). Just as the Lower and Middle Voltaian sediments have their correlatives in the Togo belt, so the sediments of Supergroups 1 and 2 in the Taoudeni Basin (Table 9.1) have their counterparts in the Pharusian belt, the eastern part of the Gourma, and the Rokelide–Mauritanide belt. The continental sandstones at the top of Supergroup 2 in the eastern half of the basin are generally reddish and are probably equivalent to the Pharusian 'série pourprée', which represents a change to continental molasse-type sedimentation (Sec. 6.3.6).

Sediments in the southern part of the Taoudeni Basin are of fluviatile–deltaic and marine origin and range in age from Upper Proterozoic to Devonian (Fig. 9.1). They are generally flat-lying with a total thickness of between about 2000 and 4000 m. The overall dip is towards the middle of the basin, but in the western half in particular, as the Mauritanide belt is approached, there is some gentle folding. Broad open folds plunge gently towards the north and disappear beneath alluvium and desert sands of

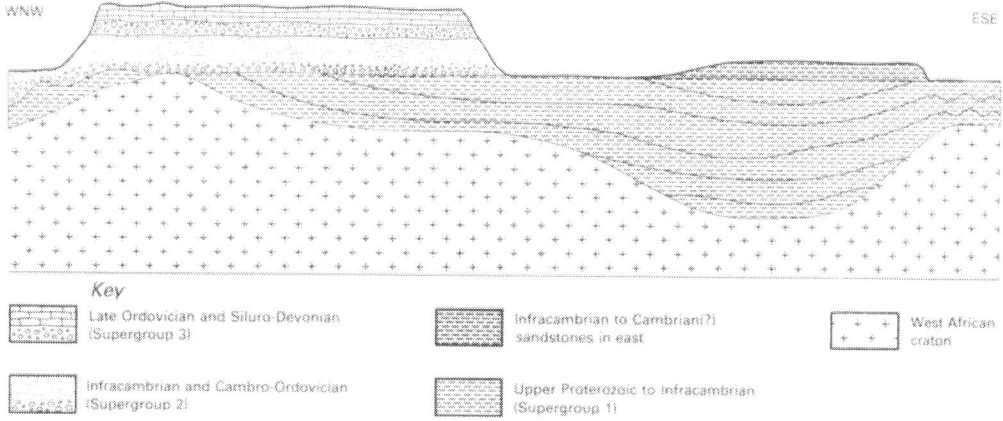

WNW   ESE

Key

	Late Ordovician and Siluro-Devonian (Supergroup 3)
	Infracambrian and Cambro-Ordovician (Supergroup 2)
	Infracambrian to Cambrian(?) sandstones in east
	Upper Proterozoic to Infracambrian (Supergroup 1)
	West African craton

Figure 9.2 Generalised and schematic cross section (not to scale) for the southern Taoudeni Basin.

the Hodh. There is also folding in the Gourma, as the strongly deformed rocks of the eastern Pan African domain are approached.

Figure 9.2 is a generalised cross section for the southern Taoudeni Basin. It is intended more to illustrate the main features of the succession than to reflect the distribution of outcrops. Infracambrian to Lower Palaeozoic sediments are thicker in the western half of the basin. Older sediments are thicker in the east, where the **Bobo-Dioulasso embayment** is a southern extension of the Gourma aulacogen (Fig. 6.3).

The sequence in the western half of the Taoudeni Basin begins with sediments of **Supergroup 1** (Table 9.1). Some of the sandstones contain tabular or vermiform trace fossils, believed to be the burrows of marine worms assigned to the genus *Scolithus*. Sediments of this part of the succession are exposed in the Affolé and Manding Plateau regions (Fig. 9.1.). They are lateral equivalents of the Bakel Group in the Mauritanides (Fig. 5.1, line A) and they are also equivalent to the Madina-Kouta Group in the south-west, south of Kedougou (Fig. 5.1, line B). Their total thickness in the western part of Figure 9.1 is up to 2000 m.

Supergroup 2 begins with an unconformity, upon which lies the triad of presumed glacial sediments, limestones and cherts (Table 9.1). Its development is sporadic, however, and the thickness of individual units varies considerably. In places the sequence is incomplete or different from the 'standard' shown in Table 9.1. It is succeeded by a variety of mainly marine sediments, including greenish shales,

sandstones and greywackes, stromatolitic limestones and dolomites, as well as cherts and jaspers (silexites), and generally terminating with cross-bedded fluvio-deltaic sandstones that are often red in colour. These sediments correspond to the Falémé Group of Section 5.2 (Fig. 5.1, line B), and range in age from Infracambrian (Vendian) to Ordovician. Their total thickness in the western part of the basin is up to about 1000 m.

Supergroup 3 also begins with sediments of glacial and fluvioglacial origin (the **Tichit Group**) that rest disconformably on older sediments. They are succeeded by shales and sandstones that are fossiliferous in places, indicating a Silurian to Devonian age range. Their total thickness is only of the order of 100–200 m.

In the eastern half of the southern Taoudeni Basin, the sediments of Supergroup 1 have a broad shallow synclinal arrangement in the Bobo-Dioulasso embayment (Fig. 9.1). The basal members of the sequence in this region include both fluviatile sandstones and marine shales. The sediments thicken towards the NNE, where there are stromatolitic limestones and dolomites, along with jaspers and cherts. These sediments lie along the axis of the Gourma Trough, and they are the lateral equivalents of the strongly folded sequences in the Pan African belt at the edge of the craton east of the Gourma. Their total thickness reaches 3000–4000 m, considerably greater than sediments of corresponding age in the west (Fig. 9.2). They were gently folded in the Pan African, being protected from major deformation by their position upon the craton itself. They are unconformably overlain by conglomerates and the massive, cross-bedded and often silicified **Bandiagara Sandstones**, which are of fluvio-deltaic origin and indicate generally easterly and northeasterly transport directions, i.e. towards the craton margin. This is a diachronous sequence, thought to be of late Infracambrian to Cambrian age, post-dating the Pan African event. Evidence of the late Precambrian glaciation is not seen in these eastern parts of the Taoudeni Basin. Either its influence did not extend this far, or all signs of it have been removed by erosion.

9.3.1 *Geochronology and development of the Taoudeni Basin*

There is a variety of geochronological data to support the correlations and age sequences given in Table 9.1. A Rb/Sr age of 860 ± 35 Ma has been recorded from illite in shales of the middle part of Supergroup 1. The argillaceous formations overlying the tillite in Supergroup 2 give an age of 595 ± 43 Ma. Sparse inarticulate brachiopod fossils in *Scolithus*-bearing sandstones in the upper parts of Supergroup 2 are assigned to the Cambro-Ordovician boundary.

The history of the southern Taoudeni Basin probably began with the Infracambrian transgression over the peneplained craton surface, around 1000 Ma ago. As in the Volta Basin, the surface must have been locally uneven and the transgression probably began at different times in different places. The major source of the detrital sediments of Supergroup 1 lay in the shield areas bordering the basin to north and south, for to the east and west were major depositional basins that subsequently became the Pan African mobile belts. The sediments were probably distributed by tidal currents, and the development of limestones with stromatolites is indicative of shallow marine conditions and probably of a warm climate at this time (but see Sec. 9.4). The tillites(?) at the base of Supergroup 2 are best developed in the north, and the ice sheets responsible presumably originated from there. In contrast, it appears that the mainly continental red sandstones of the upper part of Supergroup 2 were derived from the south and west, from both the uplifted craton and the adjacent Mauritanide belt. The Bandiagara Sandstones in the east are older. They are also post-Pan African, but they are derived from the craton and are not a molasse deposit from the adjacent orogenic belt.

There are lateral variations of thickness and facies in the sediments of the southern Taoudeni Basin.

For example, Infracambrian stromatolites are not found everywhere, local discontinuities have been recorded in some places, and the late Precambrian to Lower Palaeozoic sequences are very different in the east and west – the eastern half of the basin is devoid of marine Palaeozoic beds. The triad that marks the supposed Vendian glaciation is developed only in the west, and even this is not always seen in its entirety. These features suggest that conditions were not uniform throughout the basin and that rates of subsidence varied not only with time but also from place to place.

9.3.2 *The Bové Basin*

It is clear that the Hercynian event was of minimal importance in the southern part of the Rokelide–Mauritanide belt, for the Ordovician to Devonian sediments of the gently synclinal Bové Basin overlap on to the southern segments of the Rokelide–Mauritanide belt (Fig. 9.1). They also lie upon the sediments of the Falémé Group, which form a largely undeformed sequence on the cratonic block between the Rokelide and Koulountou belts. This group begins with the triad, which is overlain by shales, and represents Supergroup 2 of the Taoudeni Basin succession (Table 9.1). Older sediments of the Madina-Kouta Group (over 1500 m of sandstones, shales and limestones) lie to the east of the northern Rokelides (Fig. 5.1, line B) and represent Supergroup 1.

The Ordovician sediments consist mainly of several hundred metres of sandstones, often feldspathic and cross-bedded, along with some shales and locally a basal conglomerate. Lenticular alternations of sands and shales towards the top of the sequence are interpreted as glacial in origin, and may thus be correlated with the base of Supergroup 3 (Table 9.1). In Sierra Leone, these sediments are known as the **Saionia Scarp Group**, and further north they are called the **Youkounkoun Group**. The overlying Silurian and Devonian sediments consist of a few hundred metres of shales with some sandstones. The Silurian shales are black and pyritic and contain graptolites, whereas the Devonian sediments are characterised mainly by brachiopod fossils.

Sediments in the Bové Basin are scarcely folded, but they are cut by several faults, mainly with downthrow to the west, and they are intruded by numerous Mesozoic dolerite sills. Both these features are probably related to early stages of the break-up of Gondwanaland (Part III).

The Lower Palaeozoic sediments of the Bové

Basin and the western parts of the Taoudeni Basin all represent shelf sequences. They accumulated on a slowly subsiding but otherwise generally stable continental margin, and are the lateral equivalents of sediments deposited in deeper water west of the craton. These were subsequently deformed and uplifted in the Hercynian orogeny that built the northern part of the Mauritanide belt.

9.3.3 The Iullmedden Basin

This is one of the major Mesozoic basins in West Africa, but its development began in the Lower Palaeozoic. Cambro-Ordovician sediments overlie the basement in the embayment between the Aïr and Adrar des Iforas (Fig. 6.3). They are mainly of continental origin, though marine intercalations may also occur, and they are dominated by arenaceous quartz-rich facies. They rest on an uneven surface, and there is evidence that this region was also affected by the late Ordovician glaciation. The sediments are in part molasse deposits derived from erosion of the Hoggar block following the Pan African orogeny. They are overlain by predominantly marine sediments, including fossiliferous shales and limestones, that range in age from Silurian to Lower Carboniferous and pass up into estuarine and fluviatile Permo-Carboniferous sandstones and shales with plant remains.

There is much more extensive development of this Palaeozoic sedimentary sequence to the north and north-east of the Hoggar block. Here the sediments form the southern border of the huge Mesozoic basin of northern Algeria and eastern Libya. Here, too, the Ordovician sediments are more marine in character, and there is more widespread evidence of the glaciation event.

9.4 The glaciation events

The late Precambrian (Vendian, *c.* 650 Ma) and late Ordovician (*c.* 450 Ma) glaciations have already been mentioned several times. This section examines the evidence for them in more detail and considers some of the implications.

Vendian Unsorted bouldery deposits are known from sequences of late Precambrian age over much of West Africa, including the Taoudeni and Volta Basins (Table 9.1), the Pharusian belt of the Hoggar massif, and the Rokelide and Mauritanide belts. They are extensively developed further south, in

Angola and western Zaire, and are also known from Namibia. The identification of these deposits as tillites has been questioned by some authorities. Partly for this reason, the deposits are often called **mixtites** or **tilloids**, reflecting uncertainties about their origin.

In the Taoudeni and Volta Basins, the Vendian mixtites occur in stable platform sedimentary successions totalling not more than 3000–4000 m in thickness, in which clastic sediments are predominantly fine-grained. Fragments in the mixtites range in size from tens of centimetres up to over a metre; they are of variable shape and consist of many different rock types (including carbonate, siltstone, sandstone and quartzite), and are commonly striated. The matrix is an unbedded greenish argillaceous sandstone, and there are pebbly sandstone lenses within the bouldery deposits. Associated sediments are conglomerates, cross-bedded sandstones and shales, including finely laminated clays with pebbles, which are interpreted as glacial varved clays with **dropstones** (rock fragments dropped from floating ice as it melts). The surfaces beneath these deposits display features interpreted as striated pavements, **roches moutonnées** and patterned ground (polygonal patterns produced by alternate freezing and thawing in cold climate regions).

Taken together, these features have led to the confident identification of Vendian mixtites in the Taoudeni and Volta Basins as true tillites, i.e. as consolidated and lithified boulder clay. Those in the Angola–west Zaire region occur in a thick geosynclinal succession and have been interpreted as mudflow debris deposited along a subsiding continental margin. This interpretation cannot be so readily sustained in the Taoudeni and Volta Basins, which developed on stable cratonic crust; the equivalent mixtites in the Buem could be of this origin, but sedimentary lithologies point to a generally shallow-water environment of deposition.

An important element in this controversy is the close association of the triad (Table 9.1) with the mixtites. Similar associations are known from late Precambrian terranes in places other than West Africa, for example central Australia, so the problem is one of general interest and significance.

As carbonates (particularly dolomitic and stromatolitic beds) are generally considered to be diagnostic of warm-water environments, it is difficult to reconcile a juxtaposition of glacial and warm-water sediments. It can be argued either that mountain glaciers could have supplied debris to warm coastal areas at low latitudes, or that as the transition from cold glacial to warm interglacial episodes occurred on timescales of a few thousand years in the Pleistocene, it could well have done so in the distant past. More impressive, however, is recent evidence from Antarctica, where carbonate sediments, stromatolites and sulphate evaporites are now known to form in cold arid-climate lakes of high salinity. On the other hand, it should be noted that the occurrence of rich phosphates in shales overlying the triad (Sec. 9.6) is strong evidence of warm-water conditions. Sedimentary phosphates form mostly in shallow continental-shelf environments at low latitudes.

Palaeomagnetic evidence suggests that West Africa lay close to the South Pole in both late Precambrian and Lower Palaeozoic times, and thus supports the identification of the mixtites as true tillites. Australia also lay close to the pole in the late Precambrian, and tillites of this age are associated with triad-type sediments there.

Late Ordovician   The Ordovician Tichit Group of the Taoudeni Basin can be divided into two sequences. The lower sequence consists partly of elongate meandering cross-bedded sandstone bodies, which are interpreted as proglacial channel fill or esker deposits, and they rest with considerable erosional discordance on older beds. They are separated by argillaceous pebbly and bouldery sandstones, identified as continental tillites. The upper sequence comprises well bedded sandstones with occasional pebbles and boulders, which are thought to represent marine tillites and which pass up into the Silurian shales.

Sediments in upper parts of the Saionia Scarp Group of Sierra Leone include laminated mudstones with angular fragments interpreted as ice-rafted dropstones, and these deposits are also believed to have been formed under glacio-marine conditions. Like the upper parts of the Tichit Group, they pass upwards into Silurian shales.

These and other features are consistent with the development of a late Ordovician ice sheet, which moved generally north and west from a centre somewhere to the south of the Taoudeni Basin. Its products are only preserved where they were covered by shelf sediments deposited in the Silurian marine transgression on to the craton. However, the palaeomagnetic data suggest that the main development of the Ordovician glaciation lay to the north, in the Hoggar region and round the northern margin of the Taoudeni Basin, and this is also indicated by the

nature of the Ordovician sediments that border the Hoggar block (Sec. 9.3.3).

9.5   Palaeozoic sediments in coastal basins

The main development of the coastal basins in West Africa took place in the Mesozoic and they will be considered in more detail in later sections. However, Palaeozoic sediments are also found along the coast, from Sierra Leone and Liberia in the west, to Ghana in the east.

Evidence of Lower Palaeozoic sediments underlying the continental shelf along the coast of Sierra Leone comes from geophysical data, but it is consistent with surface outcrops and borehole records further east. There are continental sandstones of probable Devonian–Carboniferous age near Monrovia, called the **Paynesville Sandstone**, estimated to be about 1 km thick in total. On the coast of Ghana, there are shales and sandstones – in part marine – ranging from Ordovician to Lower Carboniferous in age. The Ordovician sediments show signs of deposition in a fluvioglacial environment.

These occurrences of Lower Palaeozoic sediments lie along the line of subsequent separation between Africa and South America. One possibility is that they were deposited in a more or less continuous linear basin of deposition that was a zone of subsidence representing an early phase of crustal stretching and rifting, about 200–300 Ma before the main break-up began. It is rather more likely that they were fortuitously preserved because they happened to lie along this line and were preserved by crustal sagging and downfaulting when the coastal basins began to develop in the Mesozoic. The extent of Lower Palaeozoic shelf sedimentation may thus have covered much more of the West African craton than indicated by the outcrop distribution on Figure 9.1. In particular, the Palaeozoic sediments in these coastal basins indicate a substantial marine transgression in the Devonian, over the southern part of the craton at least. This was presumably an extension of the shelf seas that covered what is now the Bové Basin. It is also possible that they were linked with the Lower Palaeozoic forerunner of the Iullmedden Basin far to the north (Sec. 9.3.3). Differential uplift of the cratonic crust following the separation of Africa and South America would have resulted in the erosion of these sediments over wide areas, destroying the evidence of such ancient inter-basin connections as may have existed.

9.6 Economic potential of the older basins

The sediments in the older basins are dominated by sandstones and shales of both continental and shallow marine origin. There seems little prospect of finding substantial hydrocarbon accumulations. Petroleum is not known to occur in any quantity in Precambrian sediments, and there are not many Palaeozoic oilfields. In any case, the Palaeozoic sediments in the southern part of West Africa are thin and restricted in extent, and are unpromising areas for oil exploration, except perhaps in offshore parts of the coastal basins. Nonetheless, exploration continues in the Taoudeni Basin and there are reports that oil shales have been found. Carboniferous sediments cover several thousand square kilometres in the northern Taoudeni Basin, and are partly of estuarine-deltaic origin. Although they are carbonaceous in places and contain plant remains, no coal seams of any thickness are known.

There are some Carboniferous coals in Niger, however, in the northern (older) part of the Iullmedden Basin. The Anou Araren deposit was opened up in 1980, with estimated reserves of about six million tonnes, sufficient to sustain an annual output of 200 000 tonnes for a substantial period to fuel a power station. Apart from this, the exploitable coals in West Africa are all in Mesozoic or younger sediments (Ch. 12).

Shallow-water marine sediments of the continental-shelf environment, however, may be enriched in *phosphate*, the principal raw material for fertilizer manufacture. At the northeastern end of the Volta Basin (Fig. 9.1), close to the boundaries of Benin, Niger and Upper Volta, there are deposits that have been estimated to contain around 100 million tonnes of phosphate ore, with grades that range between about 15% and 30% P_2O_5. The phosphate occurs in the shales that overlie the triad at the base of the Pendjari Group (Table 9.1), and is also found in the stratigraphically equivalent parts of the Buem Formation on the western edge of the Togo belt. The richest deposits occur where the erosion surface at the base of the Pendjari Group has cut down into the underlying basement. The phosphate is apatite or **collophane**, partly in grains and partly as cryptocrystalline matrix, along with quartz and clay minerals and hydrous iron oxide. There is local minor enrichment in manganese. Some of the sediments have been reworked, others retain their original sedimentary textures. There is evidence of some diagenetic recrystallisation and local enrichment,

and some grains have oolitic structure. The deposits have been surveyed and drilled, but exploitation of them is still several years away. They are relatively near to the Niger River, which offers a means of cheap transportation.

The only other economic mineral so far discovered in any quantity among the older sediments is *bauxite*. As in the Precambrian basement terranes, these deposits owe their origin to tropical weathering in the Tertiary, but they are dealt with here because the source rocks belong in the older basin successions. The largest exploited occurrences are in Guinea, where they are developed as residual deposits along a line practically on the boundary between Ordovician and Silurian shales occupying the southwestern limb of the Bové Basin (Fig. 9.1). There are three major mining and processing complexes in operation, and a fourth is projected, which will include an alumina smelting plant intended to produce eventually 150 000 tonnes of aluminium annually, using hydropower from the Konkoure River. Annual production is of the order of 10–12 million tonnes of bauxite, yielding some 3–3.5 million tonnes of alumina. Guinea ranked second only to Australia in bauxite output in 1980. In Mali, deposits of bauxite in the region west of Bamako (Fig. 9.1) are estimated to contain over 1000 million tonnes of ore, but exploitation is still some years away. Other deposits occur north-east of Bamako, where Tertiary and younger sediments overlap on to the Precambrian sediments. Bauxite has developed in places on Voltaian sediments in Ghana, but deposits are small in comparison with those already being worked in the Birimian terrane (Sec. 4.7.4). There was limited bauxite mining near Mpraeso at the southeastern end of the Kwahu Plateau during World War II.

Other economic minerals have not yet been found in any quantity. Prospecting for sedimentary *uranium* deposits has revealed some potential in southwestern Mali (east of the larger Precambrian inlier, Fig. 9.1) and also in the Gourma region. The widespread occurrence of fluvio-lacustrine and shallow marine sediments suggests that the reducing conditions favourable for sedimentary uranium enrichment might have existed in several places, in both the Volta and Taoudeni Basins – always provided that there was some primary uranium in the basement areas from which the sediments were derived. Similar considerations apply to *gold*, which has been sought in both the sediments and adjacent basement of southern Mali, south of Bamako, and occurs sporadically in the Voltaian, where occasional *diamonds* are also found. There would seem to be some potential for *barite* extracted from among the calcareous facies of the triad (Table 9.1), and economic deposits of this mineral are recorded near the base of the Middle Voltaian in Ghana.

Some of the clays in the sedimentary sequences could be used for local manufacture of bricks and ceramics, though they are likely in general to be less suitable than less compacted and indurated clays from younger sediments. Raw materials for cement are unlikely to be found in any quantity among the older sediments, for limestones are relatively scarce and often dolomitic. Here, again, there is more potential among the sediments of the younger basins.

Bibliography

Affaton, P., J. Sougy and R. Trompette 1980. The tectono-stratigraphic relationships between the Upper Precambrian and Lower Palaeozoic Volta Basin and the Pan-African Dahomeyide orogenic belt (West Africa). *Am. J. Sci.* **280**, 224–48.

Behrendt, J. C. and C. S. Woterson 1970. Aeromagnetic and gravity investigations of the coastal area and continental shelf off Liberia, West Africa, and their relation to continental drift. *Geol. Soc. Am. Bull.* **81**, 3563–74.

Black, R. 1980. Precambrian of West Africa. *Episodes* **1980**, no. 4 (Dec.), 3–8.

Black, R., R. Caby, A. Moussine-Pouchkine, R. Bayer, J. M. Bertrand, A. M. Bouillier, J. Fabre and A. Lesquer 1979. Evidence for late Precambrian plate tectonics in West Africa. *Nature* **278**, 223–6.

Cahen, L. 1982. Geochronological correlation of the late Precambrian sequences on and around the stable zones of equatorial Africa. *Precambrian Res.* **18**, 73–80.

Cahen, L. and N. J. Snelling, 1984. *Geochronology and evolution of Africa.* Oxford: Oxford University Press.

Clauer, N., R. Caby, D. Jeanette and R. Trompette 1982. Geochronology of sedimentary and metasedimentary Precambrian rocks of the West African craton. *Precambrian Res.* **18**, 58–71.

Culver, S. J. and H. R. Williams 1979. Late Precambrian and Phanerozoic geology of Sierra Leone. *J. Geol. Soc. Lond.* **136**, 605–18.

Deynoux, M. and R. Trompette 1976. Late Precambrian mixtites: glacial or non-glacial? (discussion of Schermerhorn 1974b, with reply). *Am. J. Sci.* **276**, 1302–24.

Deynoux, M., R. Trompette and N. Clauer 1978. Upper Precambrian and lowermost Palaeozoic correlations in West Africa and the western part of Central Africa. Probable diachronism of the late Precambrian tillite. *Geol. Rdsch.* **67**, 615–30.

Dia, O., J. Sougy and R. Trompette 1969. Discordances de ravinement et discordance angulaire dans 'le Cambro-Ordovicien' de la région de Meheria. *Bull. Soc. Geol. Fr.* **11**, 207–21.

Fabre, J. 1976. *Introduction à la Géologie du Sahara Algérien.* Algiers: SNED.

Furon, R. 1963. *Geology of Africa* (English edn). London: Oliver & Boyd.

Grant, N. K. 1972. Orogeny and reactivation to the west and southeast of the West African craton. In *The ocean basins and margins*, vol. 1, A. E. M. Nairn and F. G. Stehli (eds). New York: Plenum.

Lucas, J., L. Prevot and R. Trompette 1980. Petrology, mineralogy, and geochemistry of the late Precambrian phosphate deposits of Upper Volta (W. Africa). *J. Geol. Soc. Lond.* **137**, 787–92.

McElhinny, M. W. and B. J. J. Embleton 1976. Precambrian and early Palaeozoic palaeomagnetism in Australia. *Phil. Trans. R. Soc. Lond.* A **280**, 417–31.

Mensah, M. K. and W. G. Chaloner 1971. Lower Carboniferous Lycopods from Ghana. *Palaeontology* **14**, 357–69.

Notholt, A. J. G. 1980. Economic phosphatic sediments: mode of occurrence and stratigraphical distribution. *J. Geol. Soc. Lond.* **137**, 793–805.

Nyema Jones, A. E. and W. E. Stewart 1972. General geology of Liberia. In *African geology*, T. T. J. Dessauvagie and A. J. Whiteman (eds). Ibadan: Ibadan University Press.

Parker, B. C., G. M. Simmons Jr, F. G. Love, R. A. Wharton Jr and K. G. Seaburg 1981. Modern stromatolites in Antarctic dry valley lakes. *BioScience* **31**, 656–61.

Prasad, G. 1983. A review of the early Tertiary bauxite event in South America, Africa and India. *J. Afr. Earth Sci.* **1**, 305–14.

Saul, J. M., A. J. Boucot and R. M. Finks 1963. Fauna of the Accraian Series (Devonian of Ghana), including a revision of the gastropod *Plectonotus. J. Palaeont.* **37**, 1042–53.

Schermerhorn, L. J. G. 1974a. No evidence for glacial origin of late Precambrian tilloids in Angola. *Nature* **252**, 114–16.

Schermerhorn, L. J. G. 1974b. Late Precambrian mixtites: glacial and/or non-glacial? *Am. J. Sci.* **274**, 673–824.

Trompette, R., P. Affaton, F. Joulia and J. Marchand 1980. Stratigraphic and structural controls of late Precambrian phosphate deposits of the northern Volta Basin in Upper Volta, Niger and Benin, West Africa. *Econ. Geol.* **75**, 62–70.

Zimmerman, M. 1960. Nouvelle subdivision des séries ante gothlandiennes de l'Afrique occidental (Mauritanie, Soudan, Sénégal). *XXI Int. Geol. Congr. (Copenhagen)* **8**, 26–36.

10 *Mesozoic to Tertiary basins – inland basins*

STUDY NOTE

The next three chapters all relate to the Mesozoic to Tertiary basins of the southern part of West Africa. Chapter 10 deals with inland basins, Chapter 11 with the Benue Trough and coastal basins and Chapter 12 with the economic potential of all of them. This is a rather artificial subdivision and there is inevitably some overlap between chapters; but the alternative would be a single chapter of excessive length.

SUMMARY

After the seas retreated from Africa at the end of the Carboniferous there was a long period of continental erosion and sedimentation. The products of this phase did not begin to accumulate to any significant extent until the later Jurassic and Lower Cretaceous.

A major marine transgression in the Upper Cretaceous filled large shallow basins in the non-cratonic (Pan African) areas of the continental interior. Sea level fluctuated and shorelines advanced and retreated during the Upper Cretaceous. The maximum advance was in the Turonian, and the final regression took place in the Palaeocene, since which most of Africa has been above sea level. Continental sedimentation continued through the Tertiary in the inland basins, and there is a substantial thickness of Quaternary sediments in the Chad Basin.

Cretaceous marine sedimentation was of shallow-water type, depositing limestones, sandstones and shales, with lithologies and fossils that indicate generally warm climatic conditions throughout. Maximum sediment thicknesses in the lullmedden and Chad Basins are of the order of 2000–3000 m.

10.1 *Interregnum:* Upper Palaeozoic to Mesozoic

In the northern half of Africa, the Carboniferous was a period of marine regression, followed by deposition of fluvio-deltaic sediments with plant remains. There was a major glaciation in southern Africa during the Carboniferous, but its effects did not extend as far north as West Africa. The Hercynian orogenic movements that built most of the Mauritanides also uplifted and gently folded the Palaeozoic sediments of the Taoudeni Basin, as well as modifying and enhancing the differential uplift of the 'swells' that surround it.

Igneous activity broke out in several places during the Triassic and Jurassic, heralding the break-up of Gondwanaland (Part III). Permo-Triassic to Jurassic sediments are well developed in marginal basins in southern and eastern Africa, which formed along the zones of lithospheric extension where the continental masses later separated. Elsewhere, however, sediments of this age are sparse and dominated by continental facies (except in the Atlas mountains which are part of the Alpine chain), because most of Africa was above sea level.

Continental sedimentation continued into the Lower Cretaceous and became more widespread. The final break-up of Gondwanaland occurred in the Cretaceous, with the formation of the Atlantic Ocean, and was accompanied by global marine transgressions. During the Upper Cretaceous (Cenomanian) to Palaeocene interval, shallow seas spread across large areas of Africa, especially northwestern parts, and marine sediments were deposited in the developing Mesozoic basins. Sediments also accumulated in the coastal basins along the subsiding continental margins. The Mesozoic basins developed mainly in non-cratonic areas, that is, on or adjacent to the more recently stabilised parts of the shield, the Pan African and Hercynian domains.

The continental sediments representing the whole of the late Carboniferous to Lower Cretaceous interval can be referred to the **Continental Intercalaire**, which thus covers more than 150 Ma. However, continental sedimentation did not become widespread before late Jurassic to early Cretaceous times, at least so far as the northern half of Africa is concerned. Accordingly, the Continental Intercalaire generally approximates to the uppermost Jurassic and the lower half of the Cretaceous, and sometimes represents only the Lower Cretaceous. Cross-bedded fluviatile and lacustrine sands and gravels, with clay intercalations, characterise the more extensive parts of this unit. The general similarity of tropical flora and mainly vertebrate fauna (fish, dinosaurs, crocodiles) found in it testify to uniform and generally warm climatic conditions over much of the continent.

In the southern part of West Africa, marine sedimentation that followed the Continental Intercalaire occurred mainly in the Iullmedden and Chad Basins in the east, with connections to the large coastal Niger Delta Basin to the south, and in the Senegal Basin in the west.

Although the Continental Intercalaire covers substantial areas of the eastern Taoudeni Basin, much of it is obscured by desert sands, especially in the south. A narrow strip of gently northeasterly dipping sands and grits borders the Hodh on the northeast, unconformably overlying the Palaeozoic sediments (Fig. 9.1). It is probably mostly of Lower Cretaceous age, though fossil wood remains (*Brachyoxylon*) suggest that it may range down to

Jurassic. Cretaceous marine sediments did not extend far into the Taoudeni Basin, being confined in this region only to a relatively narrow strip west of the Hoggar massif.

The name **Continental Terminal** is given to the continental sediments of late or post-Eocene to pre-Quaternary age that overlie the marine sediments in the continental basins. At the end of the Cretaceous the African continent experienced its last major phase of uplift and gentle warping, as it began to collide with the European continental plate and the formation of the Alpine chain was initiated. Africa has remained above sea level since early in the Tertiary, and sediments of the Continental Terminal belong to this period.

10.2 The Iullmedden Basin

The succession in the Iullmedden Basin commences with Cambro-Ordovician to Carboniferous sediments of both marine and continental facies, often highly fossiliferous (Sec. 9.3.3). These accumulated in a post-Pan African pre-Hercynian depression that can be regarded as a Palaeozoic precursor of the Iullmedden Basin and now underlies the northern part of the much larger Mesozoic basin (Fig. 10.1).

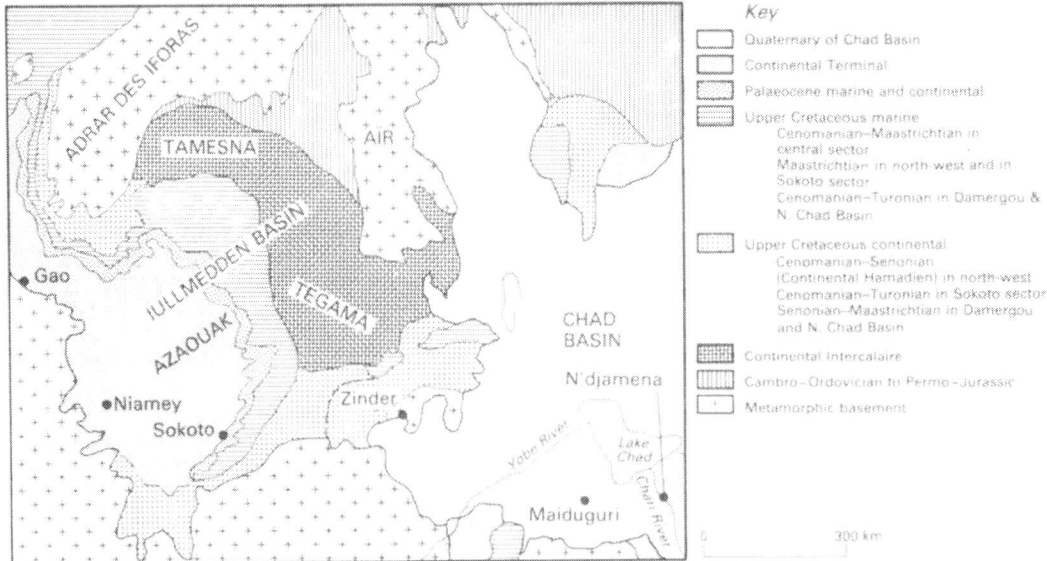

Figure 10.1 Generalised geological map for the Iullmedden Basin and the western half of the Chad Basin. Quaternary sands and alluvium in the Iullmedden Basin are not shown.

The main part of this older sequence occupies what is sometimes called the **Talach depression** or **Tin Serririne syncline**. The sediments were gently folded in the Hercynian and are overlain unconformably by the Continental Intercalaire, which here includes some Permo-Triassic to Jurassic sediments, but is dominated by the Lower Cretaceous, especially the scarp-forming sandstones of the **Tegama Sandstone Group**.

The basement blocks of the Aïr and Adrar des Iforas projecting south from the main Hoggar massif, emphasise the generally circular shape of the Iullmedden Basin (Fig. 10.1). The deepest parts of the basin are in the north-west, between the Azaouak and Tasmena regions. Here the total thickness of Mesozoic and younger sediments exceeds 2000 m. The only other part where sediment thicknesses exceed 1000 m is in the **Damergou gap** (between the Zinder and Aïr basement blocks), which links the Iullmedden Basin with the Chad Basin to the east. In this region the sediments have buried a fairly rugged topography, but in the main part of the basin they probably lie on a more or less peneplained basement surface.

The name **Sudanese Strait** (**Détroit Soudanais**) has been given to the narrow Cretaceous belt west of the Adrar des Iforas, linking the Iullmedden Basin with others further north. The **Gao Trough** is a NW–SE trending graben within the Sudanese Strait,

over 1000 m in depth, between 35 and 100 km across, and about 400 km long.

The arcuate outcrops signifying an overall inward (synclinal) dip and the southwesterly progression in age of the sediments outcropping within the basin are illustrated in the schematic cross section, Figure 10.2. An outline of the stratigraphic succession in different parts of the basin is presented in Table 10.1.

The two lowermost units of the Continental Intercalaire (Table 10.1) have a relatively restricted outcrop west of the Aïr massif. Their possible Permian–Jurassic age is based only on sparse silicified plant remains (*Dadoxylon*). The overlying **Irhazer Clays** and Tegama Sandstones are much more extensive, forming the plains of the Tamesna and the Tegama plateau respectively, which are separated by the low discontinuous Tiguedi escarpment (Fig. 10.2). In addition to silicified wood, they contain non-marine bivalve shells, fish teeth, and bones of other vertebrates such as crocodiles and sauropods, all of which combine to suggest a late Jurassic to Lower Cretaceous age.

Table 10.1 and Figure 10.1 show that the complete succession of marine Upper Cretaceous to Palaeocene strata is exposed only in the central part of the basin, in eastern Mali and southwestern Niger. This is a region of irregular topography characterised by great wadis (dry valleys) and a

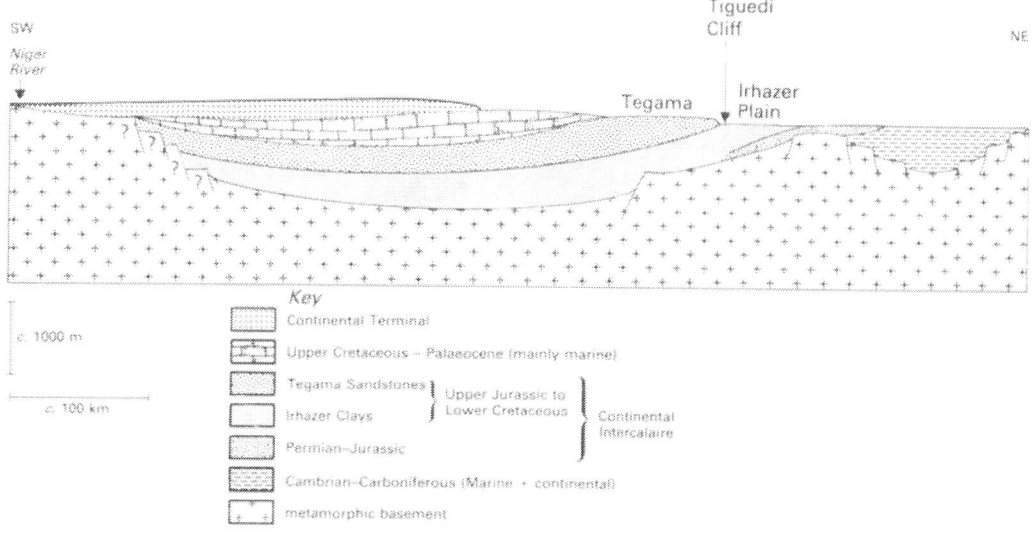

Figure 10.2 Generalised NE–SW cross section for the Iullmedden Basin. Note the greatly exaggerated vertical scale.

Table 10.1 Correlation of Mesozoic–Tertiary strata in the Iullemmeden Basin*.

Period		North-west sector	Central sector	Damergou and western Chad Basin	Sokoto sector		
Eocene–Mio-Pliocene	(Continental terminal)	Fluvio-lacustrine sands and clays with lignite fragments and oolitic ironstones (up to 450 m)		local	Gwandu Formation	(c. 300 m)	
Palaeocene–Lower Eocene		limestones, laminated ('paper') shales and marls (up to 60 m)		fluviatile and lacustrine sands and clays, locally with pisolitic bauxite (Koutous Formation in part)	Kalambaina and Gamba Formations	c. 30 m	Sokoto Group (marine)
					Dange Formation	c. 25 m	
Maastrichtian		sandstones, shales and limestones with *Libycoceras* (up to 220 m)			Wurno Formation (non-marine)	c. 50 m	Rima Group
			Marine and continental limestones and shales (up to 60 m)		Dukamaje Formation (marine)	0–25 m	
					local disconformity		
Senonian		sands and clays of the Continental Hamadien			Taloka Formation (non-marine)	c. 200 m	
					~ ?		
Turonian			Marine limestones, shales, sandstones and dolomites (up to 275 m)				
U. Cenomanian					Gundumi and Illo Formations (non-marine) (Kandi Sands in Benin)	(200–300 m)	
L. Cretaceous	(Continental Intercalaire)	Tegama Sandstone Group (up to 500 m)	Tegama Sandstone Group (up to 1000 m)	Tegama Sandstone Group			
		Irhazer Clays (up to 500 m)					
Permian–Jurassic(?)	(Continental Intercalaire)	↑?	fluvio-lacustrine sandstones and arkoses (up to 300 m)	↑?	↑?		
Palaeozoic and Precambrian			Cambro-Ordovician to Carboniferous				
		metamorphic basement	metamorphic basement	metamorphic basement	metamorphic basement		

* Not all the formations named are discussed in the text.

succession of limestone plateaux up to 100 m high, with sandy and shaly intercalations and dipping very gently southwards. The beds contain a varied molluscan fauna, including ammonites (e.g. *Nigericeras*) and the belemnite *Neolobites*, which defines the base as late Cenomanian. The fibrous clay mineral **attapulgite** is a common constituent of the shales.

In the northwestern part of the basin, round the Adrar des Iforas, the Cenomanian to Senonian stages of the Cretaceous are represented by the **Continental Hamadien** consisting of non-marine sands and clays (Table 10.1). Only the Maastrichtian and Palaeocene–Eocene are marine in this region, and there are phosphate deposits in the upper part of the succession in the Sudanese Strait (Sec. 12.2.1).

In the Damergou gap, the Continental Intercalaire is represented only by the Tegama Sandstone Group, which disappears eastwards beneath Quaternary sediments of the western Chad Basin. Marine sediments in this region contain fossils (e.g. the ammonite *Hoplitoides*) that show them to span the Cenomanian-Turonian time interval (Table 10.1), though they may range up to lowermost Senonian (Coniacian). They are overlain by late Cretaceous to lowermost Tertiary continental sediments (including the **Koutous Formation** of southern Niger), and small outcrops of Continental Terminal have also been recorded in the western Chad Basin.

10.2.1 The Sokoto sector

The southeastern part of the Iullmedden Basin extends into northern Benin, southern Niger and northwestern Nigeria, where it is sometimes called the **Sokoto Basin**, and where the sedimentology and palaeontology have probably been studied in more detail than elsewhere. The regional dip is generally north of west, but it is very small and the sediments are virtually flat-lying. They offer different degrees of resistance to weathering and erosion, so the topography of the Sokoto sector is one of ridges and shallow valleys and flat-topped hills.

The succession begins with poorly exposed fluviatile and lacustrine sediments, mainly sands and clays, with a locally well-developed basal conglomerate. These are called the **Gundumi Formation** in the north-east of the sector and the **Illo Formation** in the south-west; the formations are considered to be laterally equivalent. Towards the top of the Illo Formation there is a discontinuous horizon, up to 8 m thick, of pisolitic and nodular kaolinitic clays that are pure enough to have some economic potential (Sec. 12.2.2).

The marine Maastrichtian **Dukamaje Formation** outcrops only in a small area north-east of Sokoto and wedges out to the south and west. Where it is absent, the non-marine formations above and below it cannot be distinguished, because both are dominated by siltstones and contain no diagnostic fossils (and in consequence the wholly non-marine sediments of the Illo Formation in the extreme south of the Sokoto sector may in places be as young as Maastrichtian).

The shales of the Dukamaje Formation are often gypsiferous and contain fish and other vertebrate remains, especially turtles and crocodiles, including the genus *Mosasaurus* (hence the original name of the formation: **Mosasaurus Shales**). The formation is notable for a thin but persistent bone bed at the base, in which teeth of fish and sharks, fragments of turtle and primitive crocodiles, and coprolites are concentrated. It is a type of lag deposit, probably formed in shallow waters where currents and waves winnowed away the finer sediment particles. Other fossils include the ammonite *Libycoceras*, which confirms the Maastrichtian age of the Dukamaje Formation.

Shales of the Palaeocene **Dange Formation** are gypsiferous and contain phosphate nodules. This formation is famous for its wealth of vertebrate remains, among which crocodiles are perhaps the most interesting, as they belong to one of the few extinct reptile groups that spanned the Cretaceous–Tertiary 'crisis' during which the dinosaurs died out.

The succeeding **Kalambaina** and **Gamba Formations** are dominated by argillaceous limestones and laminated ('paper') shales respectively. The limestones have been exploited for cement manufacture, both near Sokoto and in southern Niger, where they are somewhat purer (Sec. 12.2.2). Invertebrate fossils are particularly abundant in the Kalambaina limestones and marls, and indicate a Palaeocene to Lower Eocene age for the sediments. A phosphate-rich horizon within the shales of the Gamba Formation invites comparison with the phosphatic sediments of equivalent age far to the north-west in the Sudanese Strait, where they have some economic potential (Sec. 12.2.1).

10.2.2 The Continental Terminal

In the southern half of the Iullmedden Basin, Lower Tertiary marine sediments are overlain by a thick series of ferruginised and argillaceous sandstones, mudstones and carbonaceous layers, deposited under fluviatile and lacustrine conditions – the Continental Terminal. In some places there is evidence of an erosional break at the boundary of the Continental Terminal with underlying marine sediments. In the west, the continental sediments have overlapped on to basement and they cover substantial areas of Palaeozoic strata in the southern Taoudeni Basin. In the south, Continental Terminal lies directly on the Illo Formation (Kandi Sands of northern Benin) (Fig. 10.1).

Oolitic and pisolitic ironstones occur within the sequence. The most important horizons are found locally at or near the base. They have a partly calcareous cement and contain ferruginised marine bivalves and echinoids, which suggest that they formed in shallow, warm, current-swept waters, as the sea retreated during the Eocene. Their present composition is mainly limonite and goethite with some haematite, but they may originally have been chamosite that was subsequently oxidised by diagenetic or weathering reactions. These basal ironstone layers attain a thickness of 5 m in places and their purity has attracted attention as a possible economic source of iron ore (Sec. 12.2.2). Oolitic ironstones at higher levels in the Continental Terminal were presumably deposited under non-marine conditions, but their origin is less certain and they are in any case of lesser importance.

The only fossils in the sands and clays of the Continental Terminal are plant remains that are sometimes concentrated into carbonaceous and even peaty (lignitic) layers. Analysis of pollen and spores shows that they belong to plants in the Eocene–Miocene age range. These include varieties of palm and mangrove trees, confirming deductions from the sediments themselves that conditions in the Iullmedden Basin were riverine and swampy tropical forest during deposition of the Continental Terminal.

It is relatively easy to position the base of the Continental Terminal in the Eocene, as it follows the retreat of the Palaeocene sea. It is bound to be diachronous, that is, of different ages in different places, but as the regression was probably rapid, this does not present major difficulties of correlation. The top of the Continental Terminal cannot be so easily defined, however, for it must be represented by continental deposits of widely differing ages, depending on local climatic and tectonic conditions. The most reasonable solution, in the light of presently available information, is therefore the one that has been adopted by international agreement. This is to refer all continental sediments of late or post-Eocene to pre-Quaternary age to the Continental Terminal.

In many places the Continental Terminal encompasses the lateritic capping or ferruginous duricrust that is such a common feature of West African landscapes (and of those elsewhere in Africa, too), for some of it is pre-Quaternary (Part IV). Laterite occurs indifferently on rocks of all ages where they have been subjected to deep weathering under warm humid climatic conditions. Lateritisation has been in progress throughout the Tertiary and Quaternary and locally continues today; and laterites are notoriously difficult to date. Together with alluvium and wind-blown deposits, laterite is responsible for obscuring much of the 'solid' geology, that is to say, Lower Tertiary and older rocks, including the metamorphic basement.

10.3 The Chad Basin

The boundary between the Chad and Iullmedden Basins can be conveniently taken as a roughly north–south line between the Aïr and Zinder massifs, which is the Damergou gap (Fig. 10.1). The Chad Basin is considerably larger than the Iullmedden Basin, over 1000 km in diameter, and most of it is occupied by Quaternary sands and alluvium (the **Chad Formation**) that obscure older sediments over wide areas. It is bordered by both ancient basement rocks and by Lower Palaeozoic sediments, and is a basin of internal drainage, that is to say, no rivers flow out of it. The watershed that defines the present-day limit is delineated in Figure 10.3, along with the boundary of the lake in Quaternary times, when it was vastly greater than it is now, and spilled over into the Benue River.

The present Lake Chad lies at about 240 m above sea level, but this is not the lowest part of the basin at the present time: the Bodele depression to the north-east is about 200 m lower. The Chari is the main river entering the present lake.

Information from boreholes and geophysical measurements reveals that the deepest parts of the Chad Basin are two elongate depressions. One extends north-east from the Benue Trough, beneath Lake Chad and on towards the northeastern boundary of the basin. The other extends north-west from beneath Lake Chad, to underlie the Tenere rift structure bordering the eastern side of the Aïr massif (Fig. 10.1). A third elongate depression underlies the southern part of the basin. It has an east to east–northeast trend and it lies along a projection of the Yola arm of the Benue Trough (Sec. 11.1). Positive gravity anomalies associated with these elongate depressions suggest crustal thinning, elevation of the crust–mantle boundary beneath them, and possibly the intrusion of basic igneous rocks.

Sediment thicknesses exceed 3000 m in the deepest parts of these depressions, and at least 1000 m of that total probably consists of Chad Formation clays and sands. The history of the Chad Basin thus extends well beyond the scope of this chapter, but for completeness much of it will be considered here, and then mentioned again in Part IV.

10.3.1 Cretaceous of the Chad Basin

Cretaceous sediments are exposed only in the western half of the basin, and marine Cretaceous is found only in a few localities (Fig. 10.1 and Table 10.1).

Upper Cretaceous marine sediments occupy the northeastern end of the Benue Trough where it enters the Chad Basin (Fig. 10.1). Borehole data indicate that they thicken considerably towards the north-east. More than 400 m have been proved beneath Maiduguri (Sec. 11.1.2), underlain by some 600 m of continental sediments which may be part of the Continental Intercalaire. The marine Cretaceous could exceed 1000 m thickness in the deepest parts of the basin, but insufficient borehole data are available to determine the extent of correlation between sediments in the main part of the basin and those exposed in the west. These are summarised in Table 10.1, which also shows marine sedimentation in the western Chad basin to have ended in the Turonian, though it could have continued into the Senonian. The Maastrichtian is non-marine in the west (Table 10.1) and probably marine in the southeast (Sec. 11.1.2).

There are no marine Palaeocene sediments in the Chad Basin, and it is possible that deposits placed in the Continental Terminal here range down to the Palaeocene.

10.3.2 Quaternary of the Chad Basin

Where the Chad Formation is seen at the surface it is of Quaternary (Pleistocene) age, resting unconformably on older beds. In the south-west it overlaps on to Tertiary laterites that formed on the Continental Terminal (Fig. 10.1). Chad Formation sediments bury a varied topography and consist of fluviatile and lacustrine clays and sands, with lenses of diatomite up to a few metres thick. Mammalian bones have been recovered from the sediments, including

those of a *Hippopotamus*, but few other fossils are found, apart from the diatoms, which indicate a Lower Pleistocene age. In the deeper parts of the basin, where the Chad Formation may reach at least 1000 m, the stratigraphic succession probably extends down into the Pliocene.

Detailed information about the Chad Formation in the southwestern part of the basin is provided by numerous boreholes drilled in northeastern Nigeria to evaluate and exploit the artesian water. These show that the lithologies seen in surface outcrop extend to depths of hundreds of metres with clays as the dominant lithology. There are some thick sandstone lenses, which are the aquifers, and bands of diatomite also occur at deeper levels. The boreholes here also reveal a considerable sub-Chad Formation topography near the margin of the basin. Superimposed on the overall deepening towards the north-east there are valleys and steep slopes in the basement surface, some of them possibly due to faulting.

The sediments of the Chad Formation contain a record of the climatic changes that occurred in this part of Africa during Pliocene and Quaternary times. There is great lateral variation, the sands occurring as lenses of variable thickness among the clays, which are bluish or grey and contain small amounts of carbonaceous material. The sediments accumulated in a fluviatile and lacustrine environment, the sandy lenses probably laid down as alluvial fan deposits along lake margins in times of flood. The clays represent deposition under less turbulent conditions, either away from the lake shores or at times of low river flow. The deposition of diatomite also occurred under quiet conditions, when little detrital material entered the lakes and the fragile diatom skeletons could accumulate.

Sedimentological studies of this sediment sequence suggest that it is all of sub-aqueous origin. It has been supposed that some of the sands might be aeolian, but the grains do not appear to be well enough sorted or rounded for dune sands. For most of its later history, therefore, the Chad Basin may be envisaged as a vast plain with many rivers draining into one or more lakes, probably well vegetated and well stocked with animals. Towards the end of the Quaternary, the climate became more arid, with aeolian dune sands accumulating in parts of the northern half of the basin about 20 000–40 000 years ago. In the southwestern corner, sands near the top of the Chad Formation are poorly sorted, feldspathic and contain gravel lenses, consistent with deposition

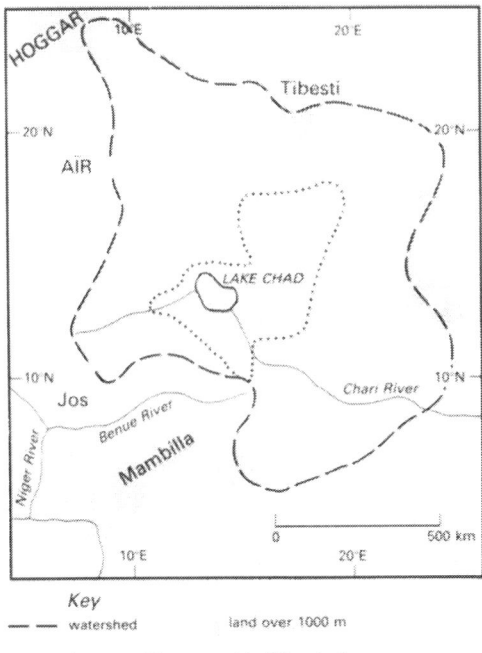

Key

— — watershed   land over 1000 m

········· boundary of Quaternary lake (Megachad)

Figure 10.3 The Chad Basin, showing the watershed (broken line), areas above 1000 m surrounding it (shaded) and the boundary of the Quaternary lake (Megachad), as defined by beach ridges (dotted line).

under relatively more arid conditions at a late stage in the history of the basin.

In the ensuing humid interval, Lake Chad reached its maximum extent (**Megachad**, Fig. 10.3) between about 9000 and 5000 years ago and lake beds are found overlying the older dunes. The boundary of Megachad is marked by a prominent beach ridge that can be traced almost all the way round the dotted line in Figure 10.3. It is a prominent NW–SE ridge passing through Maiduguri in northeastern Nigeria, where it is known as the **Bama Ridge**, and it is easily recognised in Niger and Chad also. Lower and less prominent beach ridges have been identified closer to the present lake, representing shorelines formed as Megachad shrank. This must have been an intermittent process, with oscillations of the lake shore as the overall trend to a drier climate was punctuated by alternation of arid and pluvial periods. The oscillations were linked to the waxing and waning of ice caps and glaciers in the northern and southern hemispheres during the Pleistocene ice age, that is, to alternating glacial and interglacial phases (Part IV).

10.3.3 Basin subsidence

Presently available data suggest that the thick Plio-Pleistocene succession in the Chad Basin is underlain by even thicker sediments of marine origin and Upper Cretaceous age. There is little evidence of significant Lower to mid-Tertiary sedimentation. If this is correct then, after the sea withdrew at the end of the Cretaceous, the Chad Basin must have been an area of non-deposition and remained so for almost the whole of the Tertiary, commencing to subside again only about 5 Ma ago. There should be a recognisable erosional break between the Cretaceous sediments and the Chad Formation.

The amount of subsidence need not have been great, bearing in mind that the basin is over 1000 km across and the Chad Formation sediments cannot be much more than 1 km thick. Only a small amount of relative uplift of the 'swells' round the basin would be needed to produce the height differential necessary for sedimentation to begin again. On the other hand, the *rate* of uplift might well have been considerable, to produce sufficient erosion products to accumulate up to 1000 m of sediment in the Quaternary alone. The uplifts could have been associated with the onset of late Tertiary to Quaternary volcanism in Africa, including Tibesti, Aïr and the Jos and Biu Plateaux (Part III). If the sediments date back to the Pliocene (about 5 Ma) and their maximum thickness is 1 km, then the average rate of relative subsidence of the deepest parts of the Chad Basin would have been of the order of 0.2 mm per year. As the surrounding hills were progressively eroded, the sediments encroached on to the older basement, overlapping older sediments lower down in the sequence.

Bibliography

Beaumont, C. and J. F. Sweeney 1978. Graben generation of major sedimentary basins. *Tectonophysics* **50**, 19–23.

Burke, K. 1976. The Chad Basin: an active intra-continental basin. *Tectonophysics* **36**, 197–206.

Burke, K. 1976. Neogene and Quaternary tectonics in Nigeria. In *Geology of Nigeria*, C. A. Kogbe (ed.). Lagos: Elizabethan Publishing Co.

Cratchley, G. R., P. Louis and D. E. Ajakaiye 1984. Geophysical and geological evidence for the Benue–Chad Basin Cretaceous rift valley system and its tectonic implications. *J. Afr. Earth Sci.* **2**, 140–50.

Fabre, J. 1976. *Introduction á la géologie du Sahara Algérien*. Algiers: SNED.

Falconer, J. D. 1911. *The geology and geography of northern Nigeria*. London: Macmillan.

Furon, R. 1963. *Geology of Africa*. London: Oliver & Boyd.

Kogbe, C. A. 1976. Outline of the geology of the Iullmedden Basin in north-western Nigeria. In *Geology of Nigeria*, C. A. Kogbe (ed.). Lagos: Elizabethan Publishing Co.

Kogbe, C. A. 1979. Review of the Continental Intercalaire and Continental Terminal in the Iullmedden Basin in West Africa. *Ann. Geol. Surv. Egypt* 9, 363–76.

Petters, S. W. 1981. Stratigraphy of Chad and Iullmedden basins (West Africa). *Eclog. Geol. Helvet.* 74, 139–59.

Reyment, R. A. 1965. *Aspects of the geology of Nigeria*. Ibadan: Ibadan University Press.

Reyment, R. A. 1980. Biogeography of the Saharan Cretaceous and Palaeocene epicontinental transgressions. *Cretaceous Res.* 1, 299–327.

Reyment, R. A. and E. A. Tait 1972. Biostratigraphical dating of the early history of the South Atlantic Ocean. *Phil. Trans. R. Soc. Lond. B* 264, 55–95.

Reyment, R. A., P. Bengtson and E. A. Tait 1976. Cretaceous transgressions in Nigeria and Sergipe-Alagoas (Brazil). *An. Acad. Bras. Cienc. (Supp.)*, 253–64.

11   *The Benue Trough and coastal basins*

SUMMARY

The Benue Trough is an elongate rifted depression in which the sediments reach well over 5000 m thickness in places and have been strongly folded, probably by later adjustments along faults in the underlying basement. The Bida Basin is a shallow unfaulted arm of the Benue Trough. The Benue Trough probably provided the major link between the Mediterranean (Tethys Ocean) and Gulf of Guinea via the Iullmedden and Chad Basins, during Upper Cretaceous times.

The Niger Delta at the southern end of the Benue Trough has been building out into the Atlantic since the end of the Cretaceous. Of the other coastal basins, the Senegal Basin is by far the largest, but the basins along the southern coast of West Africa are larger than their surface extents suggest, as substantial parts of their successions are offshore.

11.1   The Benue Trough

This is in many ways the most interesting of the sedimentary basins in West Africa, chiefly because of the folding movements that affected the marine and continental sediments within it. Volcanic and minor intrusive rocks are widespread and there are deposits of lead ores and coal. The trough bifurcates near its northeastern end (Fig. 11.1), and the northern branch continues beneath the Chad Formation as an elongate depression that extends well beyond Lake Chad (Sec. 10.3); while the southern branch (the Yola arm) is aligned with another deep depression beneath the southern boundary of the Chad Basin. In the northerly trending branch of the Benue Trough there is an important ridge, the **Zambuk Ridge**, represented at the surface by the basement 'spur' south-west of the volcanic Biu Plateau, and by a number of basement inliers. The sediments thin towards this ridge from both sides and there are facies changes across it. North of the ridge, the sediments are part of the Chad Basin succession, but as they have been folded they are also related to the evolution of the Benue Trough. At its southwestern end, the trough merges with the important petroleum-bearing basin of the Niger Delta, the main development of which began in the Tertiary (Sec. 11.2). The Bida Basin, now occupied by the Niger River, is a shallow late Cretaceous branch of the Benue Trough, but there is no evidence of folding or faulting of the sediments within it (Sec. 11.1.3).

Geophysical investigations (in particular, gravity measurements) have greatly assisted in determining the structure and origin of the Benue Trough. These suggest that the trough originated as a rift valley, similar to those in East Africa and elsewhere, its untypically greater width (100–120 km as against 50–60 km) being due to burial of the original boundary faults by the sedimentary fill. There is an axial zone of positive gravity anomalies, flanked by two linear negative anomalies. This arrangement is typical of rift valleys in general, and results from crustal thinning and elevation of the crust–mantle boundary beneath the central parts of the rift, possibly accompanied by the intrusion of basic igneous rocks. In the Benue Trough there is also evidence that the axial gravity high may in part be due to relatively shallow basement, an interpretation consistent with the small basement inliers found in some places (they are too small to show on Fig. 11.1). The flanking negative anomaly belts probably overlie deeper sediment-filled troughs. The axial positive anomaly disappears at about the junction with the Yola arm of the trough, but the north trending arm has an axial negative anomaly that trends northwards to the Chad Basin. The geophysical data also show that the floor of the basin is irregular and that sediment thicknesses vary considerably from place to place. A possible maximum of 6 km has been estimated for the upper Benue Trough, whereas at the lower end the maximum thickness may not much exceed 4000 m. There is no doubt that continental crust underlies

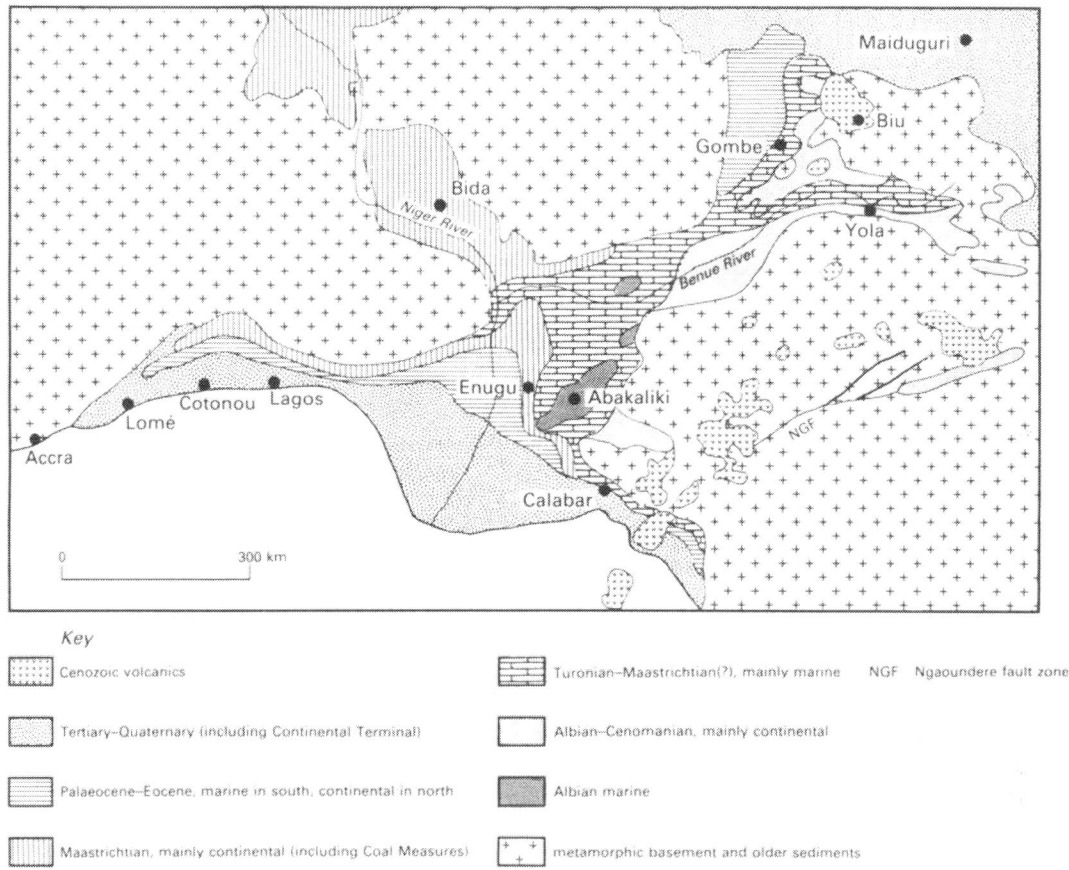

Figure 11.1 Simplified geological map of the Benue Trough, Bida Basin, Niger Delta and Dahomey Basin.

the whole of the Benue Trough, probably as far down as latitude 6°N. The irregular floor is probably the result of extensive block faulting, initiated when the trough began to develop as the 'failed arm' of a triple junction, an aulacogen that formed when Africa and South America separated in the Cretaceous. The bifurcation at the northeastern end has been identified as another triple junction, though without any crustal spreading.

Geological mapping in the Benue Trough has been concentrated at the southwestern and northern ends of the basin and rather less is known about the middle segment. Stratigraphic nomenclature changes from one end of the trough to the other, and correlations have not been unequivocally established. One possibility is summarised in Table 11.1.

This suggests that folding in the trough was diachronous: Santonian in the south-west, post-Maastrichtian in the northeast. Whether or not the correlations in Table 11.1 prove ultimately to be correct, a composite longitudinal section of Cretaceous sediments in the Benue Trough will resemble Figure 11.2.

The following sections deal with the main features of the stratigraphy and the problems of correlation. Discussion of magmatic rocks in the sequence is deferred to Part III.

11.1.1 Albian–Cenomanian
Marine sediments of this age are confined to the lower Benue Trough, mainly on its southern side. The Albian **Asu River Group** is dominated by

Table 11.1 Correlation of Cretaceous–Palaeocene strata in the Benue Trough*.

	SW Nigeria (Lower Benue)	Middle Benue	Upper Benue	NE Nigeria — Zambuk Ridge	NE Nigeria — Chad Basin
Palaeocene and younger	Imo Shale and younger beds (Sec. 11.2)	(erosion)	(erosion)	(erosion)	Chad Formation / Kerri Kerri Formation
Maastrichtian	Coal Measures { Nsukka Form. Ajali Form. Mamu Form. }	Lafia Sandstone	Lamja Sandstone	~ folding ~	Gombe Sandstone ~?~
Campanian (Senonian)	Nkporo Shales { Enugu Shales (incl. Afikpo Sst.) Owelli Sst. }	← ?	Numanha Shale†	Pindiga Formation (incl. Gulani Sandstone near top)	Fika Shales
Santonian (Senonian)	~ folding ~				
Coniacian (Senonian)	U. Awgu Shales	← ?	Sekule Formation†		
U. Turonian	L. Awgu Shales	Wukari Formation	Jessu Formation		Gongila Formation
L. Turonian	Eze Aku Shales		Dukul Formation / Yolde Formation		
Cenomanian	Odukpani Formation (SE Calabar flank only)	Makurdi and Keana Sandstones			
Albian	Asu River Group (incl Abakaliki Shales, Gboko and Arufu Limestones and Mamfe Formation Sandstone)	Awe Formation / Asu River Group	Bima, Muri and Yola Sandstones	Bima	Sandstone

* Not all the formations named are discussed in the text.
† These two formations are regarded by some workers as lateral equivalents.

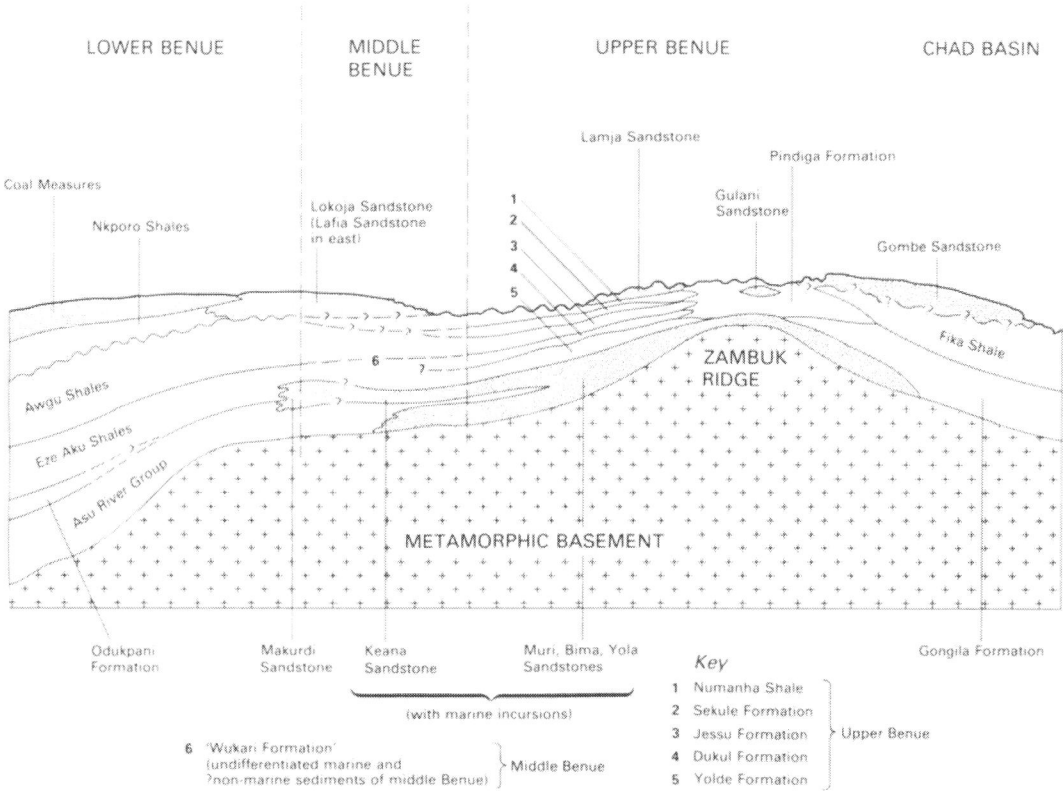

LOWER BENUE   MIDDLE   UPPER BENUE   CHAD BASIN
BENUE

Lamja Sandstone

Pindiga Formation

Coal Measures

Gulani
Sandstone

Nkporo Shales

Lokoja Sandstone
(Lafia Sandstone
in east)

Gombe Sandstone

1
2
3
4
5

Fika Shale

ZAMBUK
RIDGE

Awgu Shales

6

Eze Aku Shales

Asu River Group

METAMORPHIC BASEMENT

Odukpani
Formation

Makurdi
Sandstone

Keana
Sandstone

Muri, Bima, Yola
Sandstones

Gongila Formation

Key

1 Numanha Shale
2 Sekule Formation
3 Jessu Formation
4 Dukul Formation
5 Yolde Formation

(with marine incursions)

6 'Wukari Formation'
(undifferentiated marine and
?non-marine sediments of middle Benue) } Middle Benue

} Upper Benue

Figure 11.2 Composite longitudinal section (not to scale) along the Benue Trough, for Albian to Maastrichtian sediments.

blue-black shales, often carbonaceous and pyritic, with subordinate limestones and sandstones. A somewhat poorly oxygenated shallow-water environment of restricted circulation is indicated, alternating with more open marine conditions. Fossils are plentiful and the sediments are well dated by ammonites.

Marine Cenomanian is represented only by a very narrow strip of sandstones, shales and limestones (the **Odukpani Formation**), resting directly on the basement of the southern flank of the Calabar 'dome', and not shown on Figure 11.1.

In the middle and upper parts of the trough, Albian to Cenomanian sediments are dominated by fluvio-deltaic sandstones, variously called the **Bima**, **Yola**, **Muri**, **Keana** and **Makurdi Sandstones** in different places. They are massively cross-bedded and often ripple-marked arkosic sandstones and pebbly grits, with clay and shale lenses and a well developed basal conglomerate. The thickness is very

variable, ranging up to 3000 m in places, and there are highly altered basalt flows among the sediments. Folding and erosion of the sandstones has produced a varied topography of considerable relief in north-eastern parts of the basin.

The thickness, the massive cross bedding and the textural and mineralogical immaturity of the sandstones, in which the feldspar fragments are commonly fresh and angular, indicate rapid deposition of these deltaic sands, eroded from the rapidly rising flanks of the Benue depression. The shale and clay lenses represent local fluviatile or lacustrine conditions, and the clays are mainly kaolinitic, indicative of weathering in a tropical environment. The sandstones are diachronous, prograding in a south-westerly direction down the trough, towards the sea, and becoming younger in that direction.

Fluvio-deltaic and lacustrine sediments also occupy most of the embayment north of the Calabar 'dome' (east of Abakaliki), and are probably of

101

Albian–Cenomanian age, as are the pebbly and arkosic sediments of continental derivation that occur in fault-bounded troughs further east in Cameroun. The largest of these lies on a branch of the Ngaoundere fault zone (cf. Sec. 8.4).

Occasional wood fragments are virtually the only fossils in the non-marine Albian and Cenomanian sediments, so evidence of their age is somewhat circumstantial. However, marine bivalves in shales in the Bima Sandstone in the upper Benue Trough suggest a temporary marine incursion during Albian times, and there is evidence that the delta fronts advanced further down the trough in the late Albian (Fig. 11.2). This would be consistent with the Cenomanian regression that is indicated by the restricted extent of the Odukpani Formation.

The virtual absence of marine Cenomanian in the Benue Trough has led some workers to suggest that the Albian sediments of the Asu River Group underwent mild initial folding at this time. However, no break has yet been discerned between the marine Albian and Turonian strata in the Benue Trough. It follows that *either* the faunal ranges of the Albian and Turonian sediments in the main part of the trough should be extended to close the 'stratigraphic gap' between them *or* there was indeed a Cenomanian regression, not accompanied by folding, so that a simple disconformity exists between the Albian and Turonian strata.

11.1.2 Turonian–Maastrichtian

The Turonian was a time of major marine transgression in the Benue Trough and in other parts of northern Africa as well. Turonian marine sediments are found throughout the length of the basin, though with many local facies variations.

The Turonian–Maastrichtian history of the middle segment is less well known than that at the northeastern and southwestern ends. The sediments of most of this interval are grouped together as the **Wukari Formation**, comprising mainly marine shales, sandstones and limestones. Because of the problems of correlation the simplest course is to deal with each end of the basin separately.

The south-west Here the Turonian to early Senonian (Coniacian) interval is represented by the **Eze Aku** and **Awgu Shales**. As the names imply, these are mainly shallow-water shales and siltstones with interbedded sandstones and limestones (in places pure enough to be quarried, Sec. 12.2.2) and a variety of fossils including ammonites.

The middle Senonian (Santonian) was a period of folding and regression in the south-west, for Santonian strata are absent. The succeeding **Nkporo Shales** and their lateral equivalents unconformably overlie folded older beds. They form the lowest unit of the broad syncline that plunges gently south-west from near the Niger–Benue confluence and out beneath the Niger Delta (Fig. 11.1). The shales themselves form low ground and rarely outcrop, and information about them comes mainly from boreholes. They are of shallow-water origin, with thin beds of sandstone, shelly limestone and impersistent coals, as well as being locally gypsiferous. Fossils include the ammonite *Libycoceras afikpoensis*, which is diagnostic of the lower part of the Maastrichtian, and these sediments are presumed to span the Campanian–Maastrichtian interval. Round the southern end of the Abakaliki 'high' the shales give way to thick scarp-forming sandstones.

The **Coal Measures** in Nigeria are Maastrichtian and represent a period of non-marine sedimentation at the end of the Cretaceous. They have a total thickness of around 900 m. The lowermost 100 m or so form the base of the Enugu escarpment and consist of typical coal measures lithologies (alternations of sandstones, siltstones, mudstones and shales with concretionary siderite and marcasite) and contain at least five workable coal seams (Sec. 12.1.2). The upper part of the escarpment consists of over 400 m of barren massive cross-bedded sandstones, which are in turn succeeded by a further 300 m or so of generally less productive coal measures lithologies, with thin marine limestones near the top.

The distribution of Cretaceous sediments in the southwestern part of the Benue Trough shows that they were deposited in two basins. Prior to the Santonian folding, deposition was in the **Abakaliki Basin**, which was supplied with sediment from rivers flowing down the trough from the north-east. Uplift and folding of the sediments in this basin during the Santonian displaced the main axis of subsidence to the north-west, the **Anambra Basin**. The thickest accumulations of Nkporo Shales and Coal Measures were deposited there, thinning rapidly across the Abakaliki 'high' and remaining relatively thin on the other side of it (Fig. 11.1). During the Maastrichtian the Anambra Basin became silted up and extensive thickly vegetated swamps developed near sea level, on top of a broad delta fan built up by rivers bringing sediment down from the hinterland. This coal-forming environment was disturbed by a period of rapid fluviatile sedimentation, presumably resulting from movements of rejuvenation in adjoining upland areas. Large volumes of coarsely cross-bedded sandstones were deposited rapidly and vegetation had no chance to become established. A return to coal-forming conditions followed, and the thin marine intercalations in the upper part of the Coal Measures heralded a renewal of marine sedimentation that became more extensive in the succeeding Palaeocene and began the development of the Niger Delta (Sec. 11.2).

Sedimentological studies of the sandstones show a major point of contrast between the Albian to Turonian and Coniacian sediments of the Abakaliki Basin and the post-Santonian sediments of the Anambra Basin. Sandstones among the former are commonly feldspathic and relatively poorly sorted with a significant proportion of angular grains, that is, they are both texturally and mineralogically immature. By contrast, sandstones in the younger sediments are virtually free of feldspar, though the quartz grains are still rather angular. The inference is that there was some recycling of sediment following the Santonian folding episode, sediments deposited prior to folding being eroded to contribute to the sediments deposited after the folding and uplift. Sediments supplied to the Anambra Basin were probably also partly derived from basement areas which had undergone more prolonged weathering, resulting in breakdown of the feldspars, but not in any significant rounding of quartz grains.

The north-east In the upper Benue Trough, south of the Zambuk Ridge and in the Yola branch, the continental sandstones of the Albian–Cenomanian are overlain conformably by passage beds of the **Yolde Formation**, representing a transition to marine conditions in the Turonian. Thin alternations of shallow-water shales and sandstones are the characteristic lithology, with occasional nodular limestones that contain ammonites and other fossils. Towards the Zambuk Ridge the beds become thinner and more arenaceous and the Yolde Formation is not seen north of the ridge.

The Yolde Formation passes conformably up into a sequence of shales with thin nodular limestones and subordinate sandstones. On the Zambuk Ridge, these comprise the **Pindiga Formation**. They thicken south of the ridge and have been subdivided into five formations (Table 11.1). The shales are locally gypsiferous, and fossils include ammonites, fish, echinoids and oysters. The sandstones at the top of

the succession are of limited extent and locally contain thin coals, indicating the onset of non-marine conditions. There is a bed of oolitic phosphatic ironstones at the top of the Pindiga Formation, reminiscent of late-Palaeocene lithologies in the Iullmedden Basin (Sec. 10.2). No stratigraphic significance can be attached to this, it is merely a regressive facies that developed where conditions were appropriate.

North of the Zambuk Ridge, in the Chad Basin sector (cf. Sec. 10.3.1), the Pindiga Formation passes laterally into a sequence of limestones, sandstones and shales, the **Gongila Formation** below and the **Fika Shales** above. Both formations thicken towards the north-east, away from the ridge (Fig. 11.2). The Gongila Formation exceeds 400 m in maximum thickness, and a borehole at Maiduguri penetrated 400 m of Fika Shales without reaching the base. The sediments are very similar in both lithology and fossil content to their counterparts south of the ridge, and were also deposited under shallow-water conditions.

Still north of the Zambuk Ridge, the **Gombe Sandstone**, flaggy, sometimes cross-bedded and containing much concretionary ironstone, is of continental origin. The ironstones, which reach 2 m in thickness, may have formed as a type of 'ironpan' in very shallow water at times when clastic sedimentation was minimal. Impure coals of limited lateral extent, but up to a few metres thick, have been found in boreholes near the top of the Gombe Sandstone.

It is possible that the main folding of the underlying sediments occurred before deposition of the Gombe Sandstones, which have only been tilted gently eastwards. Indirect evidence that there may be an unconformity below the Gombe Sandstone comes from comparative sedimentological studies of the older Albian–Cenomanian fluvio-deltaic sandstones on the one hand, and the Gombe Sandstone on the other. The former are texturally immature, whereas the latter are free of feldspar and generally finer grained. As in the lower Benue Trough, these textural contrasts suggest that the Gombe Sandstone was derived in part from recycling of the older sediments after they were folded and uplifted. Moreover, grains of heavy minerals such as zircon, rutile and tourmaline are both more abundant and better rounded in the younger sandstones than in the older ones, which is also indicative of passage through a second cycle of erosion and deposition.

Unconformably overlying the gently tilted Gombe Sandstone is the **Kerri Kerri Formation**, which oversteps on to basement in the west. It is a fluviatile and lacustrine sequence of often reddish sandstones, grits and clays, with conglomerate lenses. The clays are kaolinitic and occasional lenses of quite pure clay occur, though most are sandy or silty.

The Kerri Kerri Formation has not been folded or even tilted, and it is capped by a prominent laterite, which is itself overlain by Chad Formation sediments (Sec. 10.3.2). Plant remains in impure coals indicate a Palaeocene age, and the beds have been correlated with the Continental Terminal (cf. Sec. 10.3.1). They were deposited following a further phase of uplift in the upper Benue Trough at the end of the Cretaceous, possibly part of the overall elevation that ended marine sedimentation throughout the region except in the Niger Delta. The textural and mineralogical maturity of the sediments suggests that they were partly derived from erosion of the folded Cretaceous beds to the east and south, partly from basement to the west.

The conflicting correlations Fossils in the marine sediments of the upper Benue Trough suggest that they range from Turonian to Maastrichtian without a break. Particularly significant is the ammonite *Libycoceras ismaelis*, identified in the upper part of the Pindiga Formation, which is recognised as a Maastrichtian zone fossil in northern Africa. The upper parts of the Fika Shales encountered in the Maiduguri borehole were dated as Campanian to Maastrichtian, and admittedly somewhat unreliable pollen analysis of Gombe Sandstone coals suggests a Maastrichtian age. On the face of it, therefore, the correlations set out in Table 11.1 would seem to be valid: folding in the Benue Trough was diachronous, i.e. of Santonian age in the south and late Maastrichtian in the north, and the sea *either* extended northeast from the Anambra Basin right up the trough and across the Zambuk Ridge into the Chad Basin *or* it extended south across the Zambuk Ridge from the Chad and Iullmedden Basins. The first of these alternatives appears to be ruled out by the limited extent of marine Campanian–Maastrichtian sediments in the Anambra Basin, and the second by the absence of known Maastrichtian marine sediments in the Damergou gap and western Chad Basin (Table 10.1).

An alternative that has found favour in some quarters is simply that the fossil evidence in the upper Benue Trough has been incorrectly interpreted and that the marine sediments in the upper Benue Trough and southern Chad Basin are all pre-Santonian. This is certainly a simpler solution, for it assigns a similar age to the main folding movements throughout the trough; but there is no *a priori* tectonic reason why this should be so. Whatever the truth of the matter, there is no doubt that there was also late Cretaceous or early Tertiary folding in the lower Benue, where gentle folds have affected Maastrichtian and younger sediments.

11.1.3 The Bida Basin

Also called the Niger Basin, this shallow elongate depression lies on a southeasterly projection of the Gao Trough. It is also a northwestward extension of the Anambra Basin, for the unfolded sediments in it are confined to the Campanian–Maastrichtian age range. They are thus lateral equivalents of the Coal Measures and at least part of the Nkporo Shales, and are predominantly non-marine in origin. These relationships suggest also that the non-marine Cretaceous at the southern end of the Sokoto sector in the Iullmedden Basin is locally as young as Maastrichtian.

Conglomerates and poorly sorted cross-bedded pebbly sandstones with clay horizons predominate, and are known as **Bida** or **Nupe Sandstones** in the north, **Lokoja** and **Lafia Sandstones** in the south. The clay lenses are largely kaolinitic and some are very pure. The kaolinite was derived from deep weathering of surrounding basement areas under the warm and humid climatic conditions that characterised much of the Mesozoic to early Tertiary in West Africa.

The Bida Basin is notable for the oolitic and pisolitic ironstones that occur at the top of the succession around Bida and near Lokoja, at the Niger–Benue confluence. The ironstones form flat-topped mesas and are up to 15 m thick, interbedded with clays and sandstones with plant remains, as well as shell beds with oysters and gastropods, indicating a marine influence. Most of the ironstones consist of goethite, but locally unweathered horizons near Lokoja are rich in chlorite (or chamosite) and siderite, which suggest original formation under reducing conditions. The beds contain a per cent or so of phosphorus and are similar to occurrences of oolitic ironstones recorded elsewhere in late Cretaceous–Palaeocene sequences and associated with marine regressions.

Estimates of the total sediment thickness in the Bida Basin from surface geological mapping suggest a maximum of only a few hundred metres, but gravity measurements have indicated thicknesses of 1000 m or more in some parts, and show that the floor is irregular (as in the Benue Trough). This was a very shallow basin that cannot at any time have accommodated more than the most ephemeral of shallow marine incursions. The gravity surveys have revealed an axial positive anomaly flanked by negative anomalies consistent with an underlying rift structure, but this remains to be substantiated.

11.1.4 Folding in the Benue Trough

In the south-west, folding of Asu River Group shales in the Abakaliki anticlinorium is relatively tight, with dips ranging from 5° to 80°. Further up the trough, folds are more open because of the involvement of massive fluvio-deltaic sandstones, and dips are generally not much more than 50° or so, commonly much less.

The sediments in the Benue Trough may locally total several thousand metres in thickness, but the average thickness is much less, and very small in relation to the thickness of underlying continental crust.

True scale cross sections for the Benue Trough such as the one in Figure 11.3 are unrealistic in that they show the basement surface itself warped or folded on wavelength scales of as little as 10–20 km and with amplitudes as large as 2–4 km. In a nonorogenic environment such as this, it is difficult to

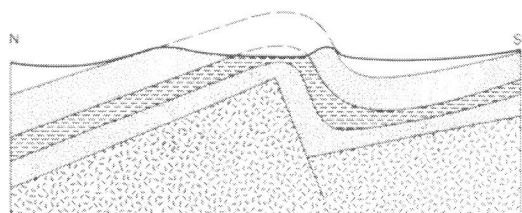

Figure 11.4 Schematic cross section illustrating how basement faulting can produce folds in overlying sediments. Note probable thickening of sediments (especially softer shales) by creep and slumping at the base of the fault scarp.

see how continental crust could be literally 'corrugated' to produce such folding. It is more realistic to envisage the movements in the basement as of block faulting type. One of the largest folds in the upper Benue Trough has straight limbs and is asymmetrical in cross section, consistent with its formation by draping layers of sediments over a fault block at shallow depth (Fig. 11.4).

The initial formation and subsequent evolution of the Benue Trough in the Cretaceous was largely controlled by rift-type faulting in the failed arm of the triple junction that developed in what is now the Gulf of Guinea. Development of the trough may have been facilitated by the more easterly grain of the basement in this part of the Pan African domain (Sec. 6.3.1). During the Upper Cretaceous, the principal movements were those of stretching and subsidence, a combination of crustal thinning and rift-type faulting, perhaps aided by sediment loading and by the emplacement of mafic intrusions inferred from recent magnetic surveys. The surface of marginal basement areas was downwarped into the basin, and the sediments have overlapped the margins, burying the main rift escarpments. The block faulting that caused the end-Cretaceous folding may have been due to compressional forces that caused reverse movements along the initial rift faults (Fig. 11.5b); or it may have been due simply to overall uplift of the whole region, readjusting the fault blocks in the basement. The second of these two alternatives seems the more likely, because the Benue Trough has the geophysical signature of a rift valley, implying thinned continental crust and an elevated crust–mantle boundary beneath it, features which are not compatible with compressional regimes. Whatever the cause, however, differential movement of basement blocks would produce drape folds in the sediments lying above. Faults have penetrated the sedimentary cover to reach the sur-

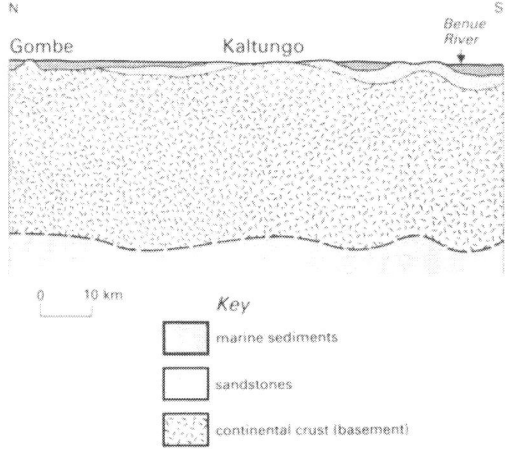

Figure 11.3 True scale cross section for part of the upper Benue Trough, showing the type of 'corrugation' of the continental crust that would be necessary to produce the folding by simply flexing or warping the basement surface.

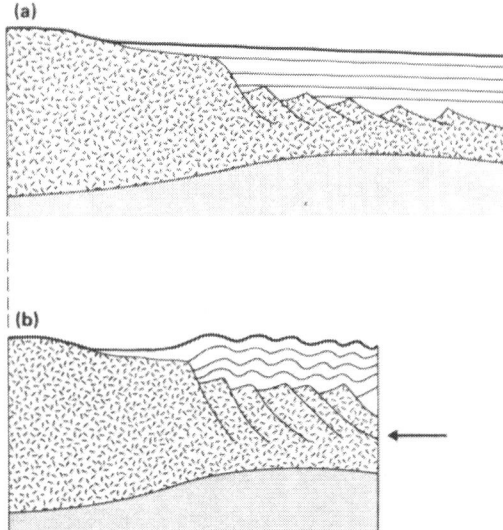

Figure 11.5 Highly schematic and vertically exaggerated cross sections to show (a) crustal stretching and normal faulting of basement to form a sedimentary basin such as the Benue Trough; and (b) crustal shortening and reverse movement along faults to produce uplift and drape folding of the sediments over the basement fault blocks.

face in several places, and there appear to be two statistically dominant directions: one more or less along the trend of the trough (NE), the other more or less across it (NW).

11.1.5 Trans-Saharan seaways

There has been some debate about Cretaceous marine links across northern Africa, between the Mediterranean in the north and the Gulf of Guinea in the south. As the mid-Atlantic ridge grew in the Cretaceous, and Africa and South America separated, the sea level rose and the continental crust of West Africa (and elsewhere) was stretched and warped. The sea transgressed southward from the ancestral Mediterranean (Tethys Ocean) across what is now the Sahara, moving along old river valleys and faulted depressions, and into the Iullmedden and Chad Basins. There were also northward transgressions up the Benue Trough.

Marine Albian sediments are virtually confined to the lower Benue Trough, representing a limited transgression; they are not known from either of the other two basins. Marine Cenomanian is virtually absent from the Benue Trough, so this was a time of regression in the Gulf of Guinea. In contrast, the Cenomanian is well represented by marine sedi-

ments in the Iullmedden and western Chad Basins. The transgression here must have come from the north, therefore, routed either to east or west (or both) of the Hoggar massif.

The eastern route, south from the Sirte Basin in what is now Libya and through the Murzuk Basin has hitherto been ruled out, because the youngest recorded sediments in the Murzuk Basin belong to the Continental Intercalaire. However, outcrops of silicified limestones of Cenomanian–Turonian age, recently reported in southernmost Libya, make this a feasible route.

A Cenomanian transgression via the Sudanese Strait and Gao Trough seems unlikely at first sight, as exposed sediments of this age are non-marine (part of the Continental Hamadien, Fig. 10.1). However, drilling in the Gao Trough beneath the Continental Terminal has recorded marine limestones of Cenomanian to Turonian age, and the Continental Intercalaire is missing. The trough must have begun to subside in the Cenomanian and Cretaceous seas probably invaded the Iullmedden Basin along it.

Marine Turonian is widespread in all three basins and this was the time of maximum transgression in the southern part of West Africa. It seems likely that they were linked by shallow open seaways, not only to one another, but also to the Mediterranean, possibly round both sides of the Hoggar block.

If Table 11.1 is correct, marine Senonian–Maastrichtian sediments in the Benue Trough are confined to the north-east (for folding brought about regression in the south-west), and these sediments extend into the southern Chad Basin at least. If Table 10.1 is correct, marine sediments of this interval are absent from the western Chad Basin. Lithologies of the Dukamaje Formation in the Sokoto sector suggest that the eastern shore line of the Maastrichtian Iullmedden 'sea' was nearby.

Both the Damergou gap and the southern Benue Trough appear to have been above sea level at this time. The Iullmedden 'sea' was presumably still linked to the Mediterranean round the western side of the Hoggar. A somewhat restricted Maastrichtian Chad 'sea', extending into the northeastern Benue Trough, may have been linked to the Mediterranean via the Murzuk Basin.

The Campanian–Maastrichtian in other parts of the world is widely recognised as a time of one of the most extensive marine transgressions in the geological record. In northern Africa it appears to have been a time of only limited transgression, if not of incipient regression.

In the succeeding Palaeocene, the sea briefly reinvaded the lower Benue Trough from the Gulf of Guinea, following the late Maastrichtian withdrawal. A Palaeocene transgression in the Iullmedden Basin penetrated well down into the Sokoto sector, but there is no marine Palaeocene in the Chad Basin. There is no field evidence of any marine connection between the Iullmedden Basin and the Gulf of Guinea during the Palaeocene. However, the ostracod faunas of this age in both regions have strong affinities, which implies that there must have been at least a short-lived Palaeocene link between the Tethys in the north and the Gulf of Guinea in the south, via the Bida Basin, evidence of which has been eroded away.

Final regression of the sea occurred during the Eocene, as a result of uplift of the continent as a whole. Marine sedimentation ended in all but the coastal basins, and continental deposits were laid down over wide areas.

11.2 The Niger Delta Basin

On the virtually flat and thickly vegetated surface of the Niger Delta plain, there are few outcrops of the belts of post-Cretaceous sediments that become progressively younger towards the south-west. They were deposited as the delta built out along the axis of the Anambra Basin (Fig. 11.1). Table 11.2 summarises the succession *seen at the surface*. It resulted from the growth of the delta into the Gulf of Guinea, following the gradual retreat of the sea after a short-lived Palaeocene transgression on to the late Cretaceous Coal Measures. The lithologies represent a variety of environments, ranging from marine through deltaic and estuarine with coastal swamps (the lignite horizons), to lagoonal and even fluvio-lacustrine. A cyclical pattern of sedimentation can sometimes be discerned, presumably the result of intermittent subsidence, leading to rapid marine incursions followed by renewed seaward advance of the deltaic deposits. Typical delta-plain sedimentary features can also be recognised among the younger deposits, including channel-fill accumulations, point bars, levées and abandoned meanders. Sedimentation in the Niger Delta Basin continues at the present time, as the Niger and Benue River systems bring large volumes of sediment to the sea each year.

The formations listed in Table 11.2 continue southwestward in the subsurface and they become progressively younger in that direction. Figure 11.6 emphasises the strongly diachronous nature of the sediments in deltas. Along the top are the formations that have been recognised and mapped at the surface. The cross section shows that they are in fact merely the *oldest* parts of sedimentary units which range up to Recent in age and which in any case represent only the upper part of the deltaic sedimentary column. The curved broken lines mark approximate successive positions of the prograding delta front, and show how the sediments become progressively younger seawards.

All deltas prograde seawards in much the same way and the pattern of sedimentation in the Niger Delta is fairly typical. River sands are dumped in numerous distributary river channels on the delta top, and are redistributed by waves and longshore currents to form beach ridges and sand bars. Muds are deposited in the quieter and deeper offshore waters over the continental shelf and slope. At the base of the continental slope, sands and muds are

Table 11.2 Palaeocene and younger sediments mapped at the surface in the Niger Delta Basin.

Miocene–Pleistocene	Benin Formation	also known as Coastal Plains Sands; cross-bedded and pebbly sands, clay lenses with lignite; marine fossils from boreholes include foraminifera, ostracods and molluscs.
Oligocene–l. Miocene	Ogwashi–Asaba Formation	clays, sands and grits with lignite seams up to more than 6 m thick
Eocene–l. Oligocene	Ameki Formation	calcareous clays, sandstones, and thin shelly limestones; some sandstones outcrop at the surface (e.g. Nanka Sand); rich in foraminifera.
Palaeocene–l. Eocene	Imo Shale	blue-grey shales with thin sandstones, marls and limestones; locally thicker sandstones in the east (e.g. Umuna, Igbalu Sst); shales contain abundant foraminifera and ostracods; correlated with Ewekoro Formation in the Nigerian sector of the Dahomey Basin (Sec. 11.3)

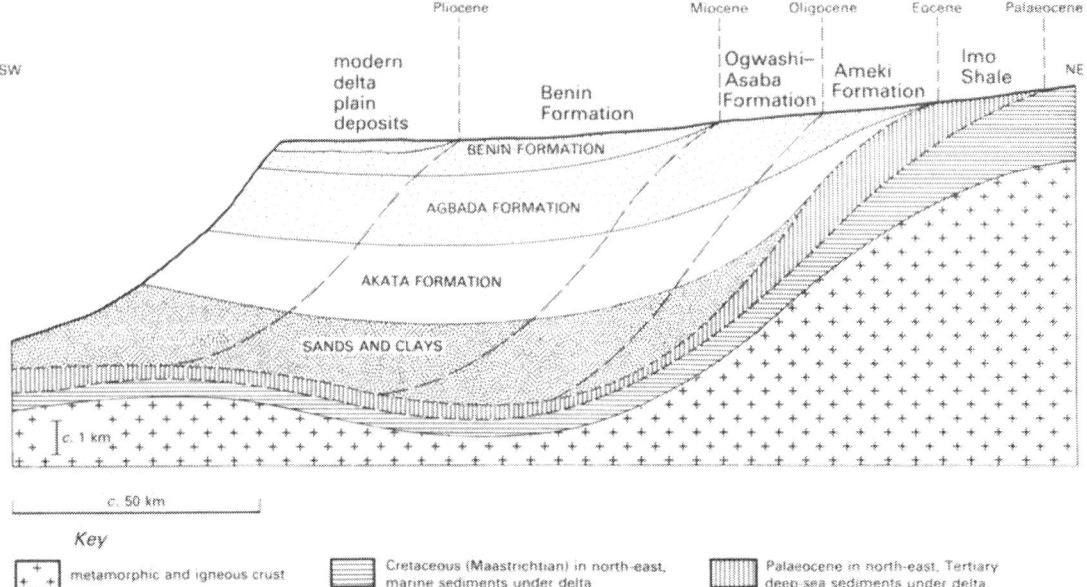

Figure 11.6 Schematic and simplified cross section of the Niger Delta Basin (not to scale), to illustrate the stratigraphy and the diachronous nature of the sediments. Curved broken lines represent successive positions of the delta front with time.

deposited in deep-sea fans from turbidity currents flowing down the slope, often in submarine canyons, and they form the continental rise. The boundaries between the formations shown on the cross section itself therefore represent only lithological changes, because the individual units are diachronous, being oldest in the north and youngest in the south.

The **Benin Formation** represents the coastal and delta-top sands and gravels, poorly sorted and often cross-bedded with clay lenses. It is up to about 2000 m thick. The **Agbada Formation** represents the immediately offshore and continental-shelf environment, consisting of alternating sands and shales above (shallower water) and mainly shales below (deeper water). Its thickness ranges from a few hundred to over 4000 m. The underlying **Akata Formation** has no surface counterpart, being the continental-slope muds (shales) and fine sands deposited on the delta front. It is between 600 and 6000 m thick. Below it are the sands and muds (clays) of the deep-sea fans forming the continental rise. The cross section in Figure 11.6 gives no hint of the structural complexity of the Niger Delta. Numerous folds and faults have been produced by slumping and differential subsidence as the thick sequences of sands and clays accumulated. The basic pattern of deformation in thick deltaic sequences is produced

by gravitational slumping. This leads to the development of normal faults (**growth faults**) towards the landward end, where tensional forces prevail. Differential subsidence along these faults produces broad anticlinal folds that provide structural traps for petroleum (Sec. 12.1.1). At the seaward end of the deltaic pile, the sediments are under compression and reverse faults develop (**toe thrusts**), as well as diapiric structures, caused here by loading of the Benin and Agbada Formations on the shales of the underlying Akata Formation. These features are summarised in Figure 11.7.

11.3 Other coastal basins

The offshore parts of the coastal basins that lie between Nigeria in the east and Guinea in the west are more important than their surface outcrops on land, for it is there that any petroleum potential is to be found (Sec. 12.1.1). The history of some of these basins commenced in the Palaeozoic (Sec. 9.5), but their main development dates from the Cretaceous.

The **Dahomey Basin** has the largest landward extent of the basins along the southern coastline of West Africa. It is bisected by the national boundary between Nigeria and Benin and extends through

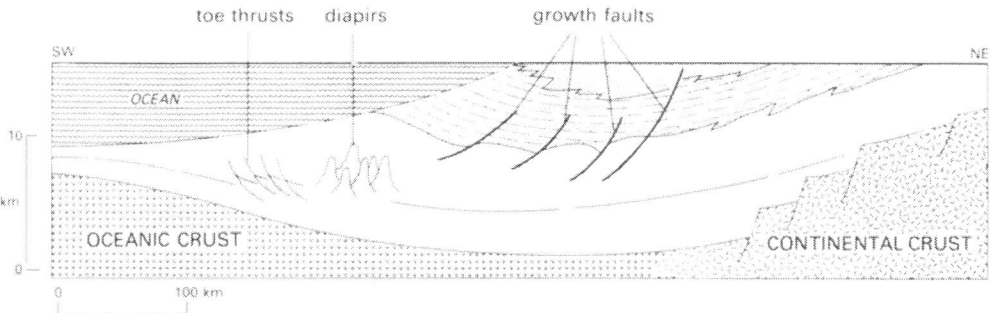

Figure 11.7 Diagrammatic and vertically exaggerated cross section to illustrate the basic pattern of deformation in unconsolidated deltaic sediments of the Niger Delta.

Togo into southeasternmost Ghana. Boreholes drilled for water and phosphate and more recently for oil exploration have helped to elucidate the stratigraphy.

At its eastern end the basin is linked to the Niger Delta Basin and Benue Trough by a narrow strip of Cretaceous and younger sediments (Fig. 11.1). At its western end lies the **Keta Basin** of Ghana, where Palaeozoic sediments are known to occur at depth (Sec. 9.5).

The Cretaceous is represented by a basal conglomerate, cross-bedded arkosic sandstones and grits with impersistent lignite and oolitic ironstone bands, and occasionally bituminous. Towards the top there are marls and clays, rich in marine fossils, indicating an upper Maastrichtian age. These sediments are called the **Abeokuta Formation** in southwestern Nigeria and they may be correlated with the Coal Measures of the lower Benue Trough and the Lokoja Sandstones of the Bida Basin. Their maximum thickness is of the order of 250–300 m. Palaeocene sediments consist of fossiliferous limestones and marls and glauconitic shales. K/Ar age measurements on specimens of glauconite from these sediments in Nigeria gave 64 ± 3 Ma, corresponding to uppermost Palaeocene. The Eocene of Nigeria (Table 11.2) comprises shallow marine phosphatic and glauconitic shales and clays with limestone lenses, as well as estuarine and deltaic sediments. Post-Eocene sands and clays are virtually all non-marine and are assigned to the Continental Terminal in Benin and Togo.

Deposition has continued to the present in offshore parts of the basins, where the thickest sequences occur. The Dahomey Basin deepens rapidly southwards, for the continental basement lies some 1800 m deep at the coast. In the west a small delta fan is building out at the mouth of the Volta River, above the Keta Basin. Boreholes here show that 3600 m of Mesozoic to Recent sediments overlie nearly 900 m of Palaeozoic marine and continental sediments (Sec. 9.5).

The post-Palaeozoic sediments in the Keta Basin are interesting, for they begin with continental sandstones and clays of Aptian and Albian age, and the earliest marine sediments are Campano-Maastrichtian. The Eocene is represented, but Oligocene sediments are absent from the Keta Basin, and there are other minor gaps in the succeeding Miocene to Recent part of the succession.

The Lower Palaeozoic sediments outcropping in places along the Ghana coast west of Accra have already been mentioned (Sec. 9.5), and they reach nearly 1000 m in total thickness. There are also over 500 m of late Jurassic to early Cretaceous conglomerates, sandstones, siltstones and shales, with freshwater crustacean and wood remains. These sediments are of terrestrial origin and can be referred to the Continental Intercalaire. They were perhaps deposited in the rifts that developed shortly before continental separation finally occurred in the Upper Cretaceous. Younger sediments are not preserved in outcrop along the Ghana coast, but they are present offshore, where some of them are petroleum-bearing (Sec. 12.1.1).

The **Ivory Coast Basin** is similar to the Dahomey Basin in some respects, though its outcrop width is somewhat less, being only 35 km at most. But here, too, much of the succession is obscured by sands and clays of the Continental Terminal, and there is a similar southward deepening, probably because of faulting. The depth to basement is over 3000 m at the coast. The eastward extension into Ghana is known as the **Tano Basin**, where the sediments are

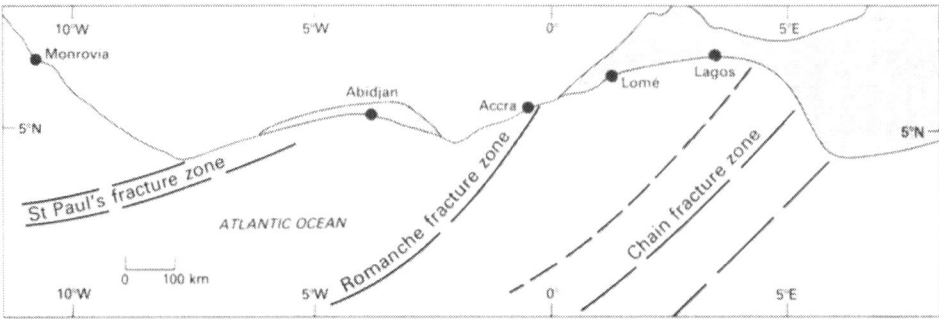

Figure 11.8 Sketch map of the southern coast of West Africa to show intersections of oceanic fracture zones with the continental margin. Sediments of the coastal basins are shaded.

referred to as the **Apollonian System**. Palaeozoic sediments are not recorded in the Ivory Coast or Tano Basins, and the lowermost Mesozoic sediments are continental Lower Cretaceous, overlain by a mainly marine sequence of limestones, sandstones and shales that range from Albian–Cenomanian to Eocene in age. There are signs of a regression in the Senonian, marked by conglomerates and bituminous sandstones, overlain locally by the Maastrichtian **Nauli Limestone**. The sequence as a whole is similar to that in the major West African basins, but differs from the coastal basins further east, where the earliest marine sediments are Campano-Maastrichtian. This could be taken to imply that the North and South Atlantic did not become fully linked until after the Senonian, but there is faunal evidence that the link was in fact established late in the Turonian, at the time of final and complete separation between Africa and South America. The absence of Turonian and lower Senonian sediments further east is therefore presumably the result of erosion or simply of non-deposition.

West of Ivory Coast, Mesozoic and younger marine sediments are confined to offshore successions that can be investigated and proved only by drilling. In Liberia the Cretaceous is represented in coastal outcrops between Monrovia and Buchanan by up to 2 km of largely terrestrial conglomerates and greywackes with plant remains, the **Farmington River Formation**. There are scattered outcrops of Tertiary sandstones as well, not more than a few tens of metres thick, again mainly terrestrial in origin, the **Edina Sandstone**. These two sequences can be correlated with the Continental Intercalaire and Continental Terminal respectively. Geophysical data indicate that up to 8 km of sediments occur off

the Liberian coast, and that there was substantial block faulting contemporaneous with the deposition of Cretaceous and younger sediments, as a result of crustal stretching and subsidence as the continents separated. The limit of continental crust may be as much as 100 km from the coast of Liberia.

In Sierra Leone, there is a coastal strip 40–50 km wide occupied by poorly exposed marine and estuarine sediments, called **the Bullom Group**. They consist of clays, in part kaolinitic, and sands, and there are lignite beds. The sediments are mainly Tertiary to Quaternary in age, as indicated by fossil fish and plant remains, and their total thickness is at least 100 m. Geophysical data indicate that there is another substantial coastal basin off Sierra Leone, with a thick Cretaceous marine sequence possibly overlying Lower Palaeozoic sediments (Sec. 9.5).

The Cretaceous and younger history of the Ivory Coast and Dahomey Basins has been influenced by oceanic fracture zones that intersect the continental margin obliquely (Fig. 11.8). As outlined in Section 8.4, there is evidence that the position of these fracture zones was originally determined by major faults and shear zones in the much older continental crust. For instance, magnetic anomalies associated with the St Paul's fracture zone system are continuous with linear anomalies in continental crust of Eburnian age. Fracture zones can also be projected into the Benue Trough and are presumably related to rift faults in the continental crust, which controlled the evolution of this basin. It must be emphasised that the predominant sense of movement along the fracture zones is vertical. The apparent 'offsets' of the southern coast of West Africa cannot be attributed to transcurrent crustal movements.

11.3.1   The Senegal Basin

The Senegal Basin is by far the largest of the coastal basins of West Africa, and in some ways also the most interesting. Unlike the other coastal basins, it is not a simple monoclinal downwarp towards the sea. Around the latitude of Dakar the basin has an overall synclinal structure, for Cretaceous and Lower Tertiary sediments are exposed there (see Fig. 9.1).

Although the total thickness of sediments is probably comparable with that beneath the Niger Delta – up to about 12 km – deposition here commenced in the Jurassic, when the North Atlantic opened (the South Atlantic did not open until the Upper Cretaceous, about 110 Ma ago). Moreover, the Senegal Basin is not developed on *c.* 500 Ma old Pan African crust, as are the other basins. Its eastern boundary is the mainly Hercynian Mauritanide belt, and most of it is probably underlain by thinned continental crust. This crust must still have been relatively warm and ductile when the North Atlantic began to open in the Jurassic, only about 150 Ma after the last collision event that ended the complex history of the Mauritanides. Geophysical data indicate that the transition to oceanic crust could be as far west as the longitude of Dakar. North–south trending growth faults controlled development of the basin and became pathways for magmas which subsequently were extruded in the Dakar region (Part III). Fold-ing and faulting in the area of the Cape Verde peninsula (the **Ndias dome**) is due to movement on the growth faults, and reflects also the presence of a basement 'swell' beneath, which was active during the Cretaceous and Tertiary (Fig. 11.9). The sediments over this 'swell' are mainly carbonates, grading laterally westwards into shales and eastwards into sandstones. Gravity measurements have also indicated the presence of large mafic intrusions in the crust beneath the basin, as might be expected in a zone of crustal extension and faulting.

Jurassic sediments are proved only in boreholes and the oldest beds exposed at the surface are Cretaceous. The largest area of these is in the south, mainly in Guinea-Bissau, at the northern end of the Bové Basin (Fig. 9.1). The sediments dip gently northwards and consist of alternations of fossiliferous shallow-water shales, sandstones and limestones, terminating with sandy Maastrichtian and amounting to a total thickness that may reach 5000 m at the coast. The succeeding Tertiary overlaps the Cretaceous beds on to older basement in places, but the sequence is relatively thin (not more than 600 m), starting with calcareous Palaeocene facies which pass up into Eocene to Miocene beds consisting of limestones, shales and marls. The sediment sequence must thicken offshore in this southern part of the basin, for there are several salt dome structures on the continental shelf, 20–50 km from the coast,

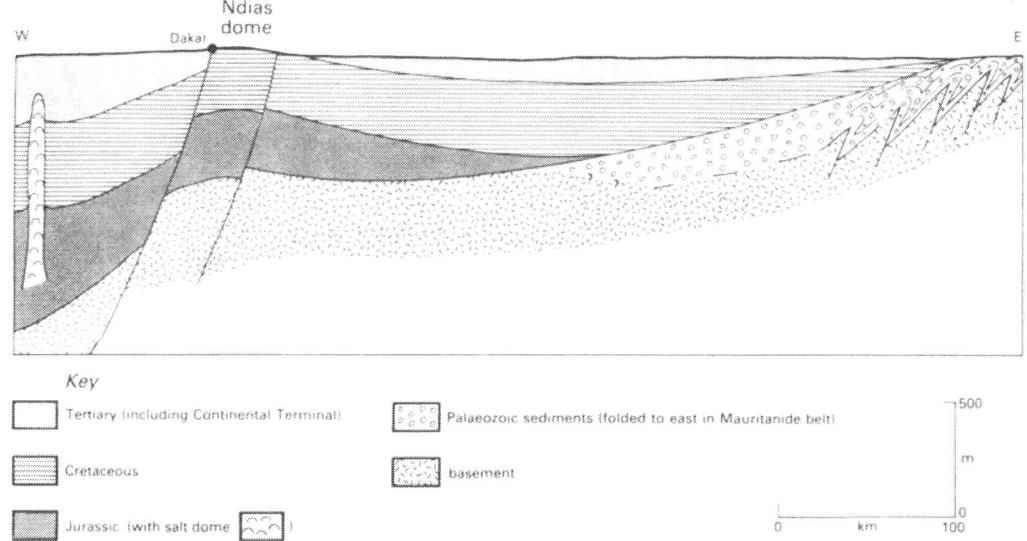

Key

Tertiary (including Continental Terminal)

Cretaceous

Jurassic (with salt dome   )

Palaeozoic sediments (folded to east in Mauritanide belt)

basement

Figure 11.9   Highly schematic cross section for the Senegal Basin, with great vertical exaggeration.

off southern Senegal, Gambia and Guinea-Bissau. These originate from evaporite deposits laid down during the early (Jurassic) stages of opening of the North Atlantic.

Further north, where the sediments have been gently folded and faulted in the Ndias dome, the oldest exposed sediments are mainly sands of Maastrichtian age. Borehole data indicate that they are underlain by a virtually complete Upper Jurassic to Senonian sequence that approaches a total thickness of 10 000 m or more. The main phase of basin subsidence ended at the close of Cretaceous times, for there is abundant evidence of a Maastrichtian regression, with development of lignite horizons in the sandy sediments. The main phase of deformation in the Ndias dome appears to have taken place at the close of the Cretaceous.

The sea advanced again in the Palaeocene, but there was a change in depositional conditions. Terrigenous sediments are relatively rare, the lithologies being dominated by oolitic and chalky limestones, with glauconitic and phosphatic horizons, cherts and attapulgite-bearing argillaceous sediments. These sediments rest unconformably on the gently deformed Maastrichtian beds, and they range up to Miocene in age, reaching some 1000 m in total thickness. The final retreat of the sea occurred at the end of the Miocene, and sedimentation of the Continental Terminal in the Senegal Basin dates from the Pliocene.

Bibliography

Adeleye, D. R. 1976. The geology of the middle Niger Basin. In *Geology of Nigeria*, C. A. Kogbe (ed.), Ch. 18. Lagos: Elizabethan Publishing Co.

Adighije, C. I. 1981. Gravity study of Lower Benue Trough, Nigeria. *Geol. Mag.* **118**, 59–67.

Adighije, C. I. 1981. A gravity interpretation of the Benue Trough, Nigeria. *Tectonophysics* **79**, 109–28.

Ajakaiye, D. E. 1981. Geophysical investigations in the Benue Trough – a review. *Earth Evol. Sci.* **1**, 110–25.

Ajakaiye, D. E. and K. Burke 1973. A Bouguer gravity map of Nigeria. *Tectonophysics* **16**, 103–15.

Akpati, B. N. 1978. Geologic structure and evolution of the Keta basin, Ghana, West Africa. *Geol. Soc. Am. Bull.* **89**, 124–32.

Anderson, M. M., A. J. Boucot and J. G. Johnson 1966. Devonian terebratulid brachiopods from the Accraian Series of Ghana. *J. Palaeont.* **40**, 1365–7.

Ani, E. I. 1980. Sukuliye (Sekulé) Formation, facies equivalent of the Numanha Shale in the Upper Benue Trough, Nigeria. *J. Mining Geol.* **17**, 91–6.

Asseez, L. O. 1976. Review of the stratigraphy, sedimentation and structure of the Niger delta. In *Geology of Nigeria*, C. A. Kogbe (ed.). Lagos: Elizabethan Publishing Co.

Aymé, J. M. 1965. The Senegal salt basin. In *Salt basins round Africa*, vol. 1, 83–90. London: Institute of Petroleum.

Beaumont, C. and J. F. Sweeney 1978. Graben generation of major sedimentary basins. *Tectonophysics* **50**, 19–23.

Behrendt, J. C. and C. S. Wotorson 1970. Aeromagnetic and gravity investigations of the coastal area and continental shelf of Liberia, West Africa, and their relation to continental drift. *Geol. Soc. Am. Bull.* **81**, 3563–74.

Behrendt, J. C., J. Schlee, J. M. Robb and M. K. Silverstein 1974. Structure of the continental margin of Liberia, West Africa. *Geol. Soc. Am. Bull.* **85**, 1143–58.

Benkhelil, M. J. 1982. Structural map of the upper Benue valley. *J. Mining Geol.* **18**, 140–51.

Buffetaut, E. and P. Taqyet 1979. An early Cretaceous terrestrial crocodilian and the opening of the South Atlantic. *Nature* **280**, 486–7.

Burke, K. 1975. Atlantic evaporites formed by evaporation of water spilled from Pacific, Tethyan and southern oceans. *Geology* November, 613–16.

Burke, K. 1976. Neogene and Quaternary tectonics in Nigeria. In *Geology of Nigeria*, C. A. Kogbe (ed.). Lagos: Elizabethan Publishing Co.

Cratchley, C. R. and G. P. Jones 1965. *An interpretation of the geology and gravity anomalies of the Benue Valley, Nigeria*. Overseas Geol. Surv., Geophys. Paper 1. London: HMSO.

Cratchley, G. R., P. Louis and D. E. Ajakaiye 1984. Geophysical and geological evidence for the Benue–Chad Basin Cretaceous rift valley system and its tectonic implication. *J. Afr. Earth Sci.* **2**, 140–50.

Culver, S. J. and H. R. Williams 1979. Late Precambrian and Phanerozoic geology of Sierra Leone. *J. Geol. Soc. Lond.* **136**, 605–18.

Falconer, J. D. 1911. *The geology and geography of northern Nigeria*. London: Macmillan.

Furon, R. 1963. *Geology of Africa*. London: Oliver & Boyd.

Hastings, D. A. and M. Bacon 1979. Geologic structure and evolution of the Keta Basin, West Africa (Discussion of Akpati, 1978, with reply). *Geol. Soc. Am. Bull.* **90**, 889–92.

Hoque, M. 1977. Petrographic differentiation of tectonically controlled Cretaceous sedimentary cycles, southeastern Nigeria. *Sed. Geol.* **17**, 235–45.

Jackson, J. A. 1980. Reactivation of basement faults and crustal shortening in orogenic belts. *Nature* **283**, 343–6.

Kogbe, C. A. 1976. The Cretaceous and Palaeogene sediments of southern Nigeria. In *Geology of Nigeria*, C. A. Kogbe (ed.). Lagos: Elizabethan Publishing Co.

Kogbe, C. A., D. E. Ajakaiye and G. Matheis 1983. Confirmation of a rift structure along the Mid-Niger Valley, Nigeria. *J. Afr. Earth Sci.* **1**, 127–32.

Lees, G. M. 1952. Foreland folding. *Q. J. Geol. Soc. Lond.* **108**, 1–34.

Lehner, P. and P. A. C. de Ruiter 1977. Structural history of Atlantic margin of Africa. *Bull. Am. Assoc. Petrol. Geol.* **61**, 961–81.

Le Marechal, A. and P. M. Vincent 1972. Le fessé Cretacé du Sud-Aclamaoua, Cameroun. In *African geology*, T. F. J. Dessauvagie and A. J. Whiteman (eds). Ibadan: Ibadan University Press.

Mensah, M. K. and W. G. Chaloner 1971. Lower Carboniferous lycopods from Ghana. *Palaeontology* **41**, 357–69.

Morel, S. W. 1979. The geology and mineral resources of Sierra Leone. *Econ. Geol.* **74**, 1563–76.

Nwachukwu, S. O. 1972. The tectonic evolution of the southern portion of the Benue Trough, Nigeria. *Geol. Mag.* **109**, 411–19.

Offodile, M. E. and R. A. Reyment 1977. Stratigraphy of the Keana–Awe area of the middle Benue region of Nigeria. *Bull. Geol. Inst. Univ. Uppsala (NS)* **7**, 37–66.

Ojo, S. N. and D. E. Ajakaiye 1976. Preliminary interpretation of gravity measurements in the middle Niger Basin area, Nigeria. In *Geology of Nigeria*, C. A. Kogbe (ed.). Lagos: Elizabethan Publishing Co.

Olade, M. A. 1975. Evolution of Nigeria's Benue Trough (aulacogen): a tectonic model. *Geol. Mag.* **112**, 575–83.

Petters, S. W. 1978. Stratigraphic evolution of the Benue Trough and its implications for the Upper Cretaceous palaeogeography of West Africa. *J. Geol.* **86**, 311–21.

Petters, S. W. 1979. Stratigraphic history of the south-central Saharan region. *Geol. Soc. Am. Bull.* **90**, 753–60.

Petters, S. W. and C. M. Ekweozor 1982. Origin of mid-Cretaceous black shales in the Benue trough, Nigeria. *Palaeogeog. Palaeoclim. Palaeoecol.* **40**, 311–19.

Reyment, R. A. 1965. *Aspects of the geology of Nigeria*. Ibadan: Ibadan University Press.

Reyment, R. A. 1980. Biogeography of the Saharan Cretaceous and Palaeocene epicontinental transgressions. *Cretaceous Res.* **1**, 299–327.

Reyment, R. A. and E. A. Tait 1972. Biostratigraphical dating of the early history of the South Atlantic Ocean. *Phil. Trans. R. Soc. Lond. B* **264**, 55–95.

Reyment, R. A., P. Bengtson and E. A. Tait 1976. Cretaceous transgressions in Nigeria and Sergipe-Alagoas (Brazil). *An. acad. bras. Cienc. (Supp.)*, 253–64.

Spengler, A. de, J. Castelain, J. Cauvin and M. Leroy 1966. Le bassin Secondaire–Tertiare du Sénégal. In *Sedimentary basins of the African coast*, part 1, D. Reyre (ed.). Paris: ASGA.

Wright, J. B. 1976. Origins of the Benue Trough – a critical review. In *Geology of Nigeria*, C. A. Kogbe (ed.). Lagos: Elizabethan Publishing Co.

12 Economic potential of the younger sedimentary basins

SUMMARY

The Mesozoic–Tertiary basins of West Africa are of great economic importance to the countries in which they occur, mainly because of the energy resources they contain. These are dominated by petroleum, but include also coal and uranium. There are large deposits of phosphate, an essential fertiliser mineral, and a host of industrial raw materials, including limestone for cement manufacture, kaolinitic clays, gypsum and salt. Metalliferous deposits are generally not so large or abundant as in basement terranes, but include ores of lead, zinc, iron and manganese, and there may also be some bauxite. Smaller deposits of these various resources are already exploited on a small scale.

12.1 Energy resources

Virtually all of West Africa's energy resources occur in the Mesozoic–Tertiary sedimentary basins. Oil and gas are of greatest importance and are dominated by the enormous reserves of the Niger Delta Basin, but there are significant accumulations in other coastal basins as well. Coal in Nigeria provided much of the energy for industrial development until well into the 1950s, and the enormous reserves will last for many decades.

Uranium deposits in Niger were discovered comparatively recently and have provided an important contribution to the national economy.

12.1.1 Petroleum

The Niger Delta is a major oil and gas province, rated by some authorities as the twelfth largest in the world. Its petroleum potential was suspected for many years, because of the numerous oil seeps in sediments of the Delta and the lower Benue Trough. Production of crude oil began late in the 1950s and grew to a maximum of more than two million barrels a day during the 1970s. In 1980, 750 million barrels were produced, making Nigeria the world's seventh largest producer in that year. Economic recession forced a reduction to little more than 460 million barrels in 1982, and just over 450 million barrels in 1983. Initially all the oil produced was exported, but with the rapid development of refining capacity (there are now four refineries), Nigeria soon became virtually self-sufficient in petroleum and petroleum products, while remaining a major exporter of crude oil – sixth in the world ranking in 1979, for example. Proven reserves of crude oil are estimated to be as much as 20 billion (10^9) barrels, sufficient for some 30 years further production at least. Utilization of the natural gas associated with the oil has not been so rapid, much of the gas being flared off, but Nigeria has great potential. It is within the world's top ten for gas reserves, at an estimated total of some 1.5×10^{12} m^3. A liquefied natural gas (LNG) plant is under construction, and some gas is already being used in the Aladja steel plant (Sec. 6.5).

More than 500 exploration and production wells have been drilled in the Niger Delta, and analysis of the data provided by them has enabled oil geologists to map the distribution of petroleum reserves. There are over 250 oil and gas fields of varying size, the great majority in the Agbada Formation (Sec. 11.2) and most of them in sediments of Eocene to Miocene age, though a few are in younger beds. The reason for this lies in the strongly diachronous nature of the Niger Delta units. The gravitational slumping along minor fault planes that characterises thick unconsolidated deltaic accumulations (Fig. 11.7) results in gentle folding, with development of anticlinal traps in the sandstones. These provide most of the reservoirs, being sealed by the clay and shale intercalations in the sequence. The shales and clays of the Akata Formation are believed to be the main source rocks for the petroleum, which migrates up fault planes into the reservoir horizons.

The belt where petroleum reserves are most abundant (Fig. 12.1) appears to be the deepest part of the

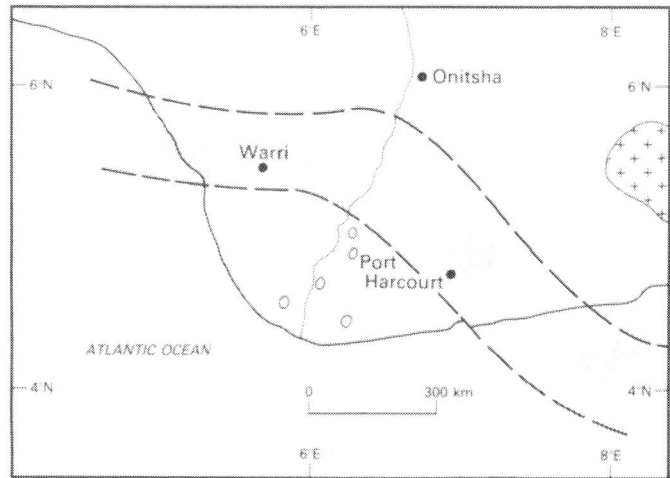

Figure 12.1 The main concentrations of reserves of oil and gas lie in five main areas (shaded) within a broad belt across the northern half of the Niger Delta Basin. Isolated smaller fields occur within the belt and also in the southern half of the delta (circles).

basin, and may overlie the transition from continental to oceanic crust in the underlying basement. The thermal gradient in this zone is of the order of 1°C per 30 m, which is significantly greater than in the southern half of the delta. The source beds must therefore lie at greater depth in the southern half of the delta, and it has been suggested that deeper drilling might intersect accumulations closer to these source beds. For example, the threshold for the commencement of oil formation from organic remains is estimated to be 115°C in Tertiary deltaic environments. At 1°C per 30 m, oil formation will commence only below depths of about 3500 m. A smaller gradient of, say, 1°C per 40 m, will mean that this critical depth is increased to 4600 m. These are minimum figures, but they serve to illustrate the point – and the chances of finding oil are statistically improved if it is sought in reservoirs that are close to the site where it has been generated.

The distribution of the five major oil field areas in Figure 12.1 may reflect a relative abundance of reservoir sandstones, and these are thought to mark the sites of ancient subdelta complexes that contributed to the overall development of the main Niger Delta Basin. If significant petroleum accumulations do occur outside the main zone of Figure 12.1 they will presumably be found in similar settings.

As the Nigeria–Cameroun border cuts across the southeastern extremity of the Niger Delta Basin, there are oil fields off the shore of Cameroun also.

Production commenced in 1977 in the coastal Rio del Rey area (close to Calabar) which now holds eight producing fields and yielded a total of 120 000 barrels per day in 1983, sufficient to supply the national refinery complex. In 1983 the new Victoria field began production, which is expected to reach 30 000 barrels per day from recoverable reserves estimated at 30×10^6 barrels. There are over 100×10^9 m^3 of natural gas reserves offshore of Kribi, and production of liquefied natural gas (LNG) could start in the 1990s.

None of the other coastal basins along the southern coast of West Africa has so far yielded promise of great petroleum potential, which is not surprising in view of their much smaller size. Nonetheless, some useful finds have been made. In the Dahomey Basin, the small Sene field, discovered in 1968 about 15 km offshore, though small by international standards, has at least 20 million barrels of recoverable reserves. Production commenced in 1982 and reached 4000 barrels per day in 1983, providing support for Benin's balance of payments, which will increase as production rises. Further west, production at the Saltpond offshore field in Ghana reached nearly 400 000 barrels in 1983, over ten times more than in 1982, and reserves are estimated at 40×10^6 barrels. Oil seeps have been known for many years in the Tano Basin at the eastern end of the larger Ivory Coast Basin, and oil and gas have been discovered in the Cretaceous sediments there. Off Ivory Coast, the Belier field began production in 1980, and the larger Espoir field was producing 18 000 barrels a day in 1982. Exploration also continues further west, off Liberia, Sierra Leone and Guinea.

115

The large Senegal Basin appears to be less rich in hydrocarbon potential than its size might suggest. The Dome–Flore oil field, about 60 km offshore from southern Senegal (where the salt domes occur, Sec. 11.3.1), is estimated to contain about 100 million tonnes of heavy oil, and up to perhaps 10 million tonnes of light oil, but exploitation has not yet begun. There is offshore exploration for both oil and gas in Gambia and both northern and southern Senegal, and there is also some potential in the coastal part of the Mauritanian segment of the Senegal Basin, around Nouakchott. Inland parts of the basin are also being explored, and a small accumulation of natural gas near Rufisque, some 60 km east of Dakar, is exploited for domestic use.

So far as the larger interior basins are concerned, the petroleum potential is more limited, for, in spite of their vast size, they are relatively shallow. An exception is the Benue Trough, where the sediment pile is certainly thick enough and the lithologies are suitable, but the degree of deformation reduces the chances of significant oil and gas finds there. Deeper parts of the Chad Basin offer some promise, and a small oil field is in production, able to provide about 70 000 barrels a day, sufficient to cover Chad's domestic demand. There are showings in southeastern Chad and northern Cameroun, and discoveries have also been made north of Lake Chad itself, in the Kanem district, as well as further west, near Tin Touma in eastern Niger, where traces of oil were found at 1500 m depth. Further small discoveries were made in the western Chad basin in 1983. The Iullmedden Basin is not a particularly promising prospect either, but there are some hopes of useful finds along the Niger River valley in the vicinity of Gao.

12.1.2 Coal

The major occurrence of coal in the younger basins is in the lower Benue Trough, where several seams occur among the Lower Coal Measures. The exploitable part of the coal field is on the gently west-dipping eastern limb of the broad synclinal structure that occupies the southwestern end of the Benue Trough (Fig. 11.1). Coals are best developed in the region round Enugu. The important seams are between 1 and 2 m thick, and there are more than 300 million tonnes of estimated reserves. The coal is a sub-bituminous variety, with a calorific value averaging around 6000 cal g^{-1} (c. 7000 kWh t^{-1}). It has a rather low carbon content of around 40%, and relatively high volatiles (c. 40%), water (up to 15%) and ash (up to 28%, but normally not much greater than 12%). The ratio of volatiles to carbon tends to increase from south to north, indicative of a slight decrease in coal rank in that direction.

Major exploitation of Nigerian coal began in 1915, with the development of rail transport in Nigeria and adjacent West African countries. Annual production rose to a maximum of over 900 000 tonnes in 1958, but has declined since, partly as a result of the increased use of oil and gas, and partly because of increased reliance on hydropower for electricity generation, both in Nigeria and in neighbouring countries. Production during the early 1980s fluctuated between about 50 000 and 100 000 tonnes annually, but there are plans for greatly increased output (up to 2–4 million tonnes per year) to meet new export demands and to fuel a projected coal-fired power station, for coal is a cheaper fuel than oil for electricity generation.

There are no other coal fields of importance in the southern part of West Africa. That is because during the Cretaceous–Tertiary there was no other suitable environment for the development of extensive coastal swamps growing on the muds and sands of a large deltaic complex accumulating and subsiding at the mouth of a major river system. Exploration boreholes have been drilled in the Lafia area, east of the Niger–Benue confluence. Thin impersistent coals of low quality occur within the estuarine sandstones and shales of the Gombe Sandstone at the northeastern end of the Benue Trough (Table 11.1). Coals have also been reported among Cretaceous sands in the coastal basins of Togo, Benin and Ivory Coast. None of these is likely to be of economic significance, however. Deposits of *lignite* occur among the Tertiary sediments of the Niger Delta Basin, notably in the Ogwashi–Asaba Formation (Table 11.2), where seams up to 6 m thick are reported, with possible reserves of some 60 million tonnes. Lignites are also known in the Iullmedden and Chad Basins, but the only other deposits of interest are impersistent lenses among the sands and clays of the Bullom Group in Sierra Leone (Sec. 11.3), especially the Koyna lignite deposit, about 25 km east of Freetown. These may be as young as Pleistocene in age. Possible reserves could reach 20 million tonnes, and the calorific value of the lignite can be upgraded nearly to that of Nigerian coal by washing and drying. The deposits are too small for anything except small-scale uses, however, such as a fuel for firing bricks made from clays of the Bullom Group.

12.1.3 Uranium

The only major uranium deposit so far located in West Africa is in Niger, where uranium mining has become a key factor in the country's economy since it began in 1971. It provides around one-third of all government revenue and about three-quarters of the total export earnings.

The main deposit is at Arlit, about 200 km NNW of Agadés at the southern tip of the Aïr, and was the first to be discovered. It occurs in Carboniferous sandstones and clays of mainly fluviatile origin in the northern Iullmedden Basin. The overlying Jurassic to Lower Cretaceous sandstones and clays of the Continental Intercalaire are also mineralised in places. The host sediments comprise lenses of coarse permeable feldspathic sandstone interbedded with clay layers. There is abundant carbonaceous debris (plant remains) in the sediments and the palaeo-environment was one of generally westward-flowing streams, draining the high ground of the Aïr massif to the east. The sandstones represent the ancient active stream channels, and the clays presumably represent alluvial deposits laid down in quieter reaches and in times of flood. There was periodic rejuvenation of the Aïr basement block, from Carboniferous times onwards, thus ensuring repeated influxes of fresh sediment into the Iullmedden Basin. The beds have been gently folded and the broad open structures are superimposed on the shallow regional westerly dip. There has also been some faulting.

The pattern of mineralisation is of secondary type, similar to that found in sediments of the western USA. It originated by the circulating of uranium-rich oxidised ground waters migrating through the porous and permeable sandstones of the palaeo-stream channels, and also along fractures. The uranium minerals were precipitated where these circulating waters encountered reducing conditions near concentrations of carbonaceous material. The main uranium mineral is coffinite, with accessory pitchblende.

The primary source of the uranium has been a subject of considerable research, and it seems most likely that the main source was the anorogenic post-Pan African ring complexes that intrude the basement of the Aïr massif and give radiometric ages ranging from Ordovician to Devonian. The complexes are dominated by granites and rhyolites (Part III), which are commonly enriched in uranium, and erosion of these rocks is thought to have provided most of the uranium in the sedimentary deposits.

At Arlit, the deposit consists of a 20–25 m thick clay and sandstone layer, beneath up to 50 m of overburden, and the ores are mined in open pits. Reserves total over 25 000 tonnes of uranium oxide, grading at 0.25% U_3O_8. On-site processing (acid leaching) is used to upgrade the ore and produce a concentrate running at 70% U. Annual production is of the order of 2000 tonnes U. Only 20 km from the Arlit deposit is the underground Akouta mine, which commenced operations in 1978, mining 0.4% grade ore at about 250 m below the surface. The ore is at the same horizon as the Arlit deposit, and production capacity is similar, except that Akouta produces *molybdenum* as a by-product. There are several other promising prospects in the Iullmedden Basin 'embayment' west of the Aïr, both in the Carboniferous and overlying continental Mesozoic sediments. These will no doubt be exploited when the price of uranium and world demand justify further expansion of mining, especially as there is a chance of producing *copper* as a by-product. Exports varied between a low of 3600 tonnes U and a high of 4700 tonnes U in the period 1980–83.

12.2 Metals and non-metals

At present, the other major economic resource of the younger sedimentary basins in West Africa is phosphate. The basins also contain most of the limestone used to make cement for the construction industry, where it is increasingly used in preference to traditional building materials. Clays for bricks and ceramics are plentiful, and other industrial raw materials include salt, gypsum and diatomite, some of which are already exploited on a local scale. Metalliferous deposits include the lead–zinc field of the Benue Trough, but these metalliferous deposits are less important. The lead–zinc ores of the Benue Trough are small and scattered, and can only be worked by small-scale mining; and although some of the Tertiary ironstones are of ore grade, they will probably not be economic propositions for the foreseeable future. The same applies to the deposits of manganese ores and bauxite which occur among the sediments in these younger basins.

12.2.1 Phosphate

The warm shallow seas of the Eocene provided an environment suitable for phosphate deposition in a number of places in West Africa. Best known are the deposits of Togo (Hahotoe, north-east of Lomé) and

Senegal (Thies and Taiba, north-east of Dakar). Both are in coastal basins, deposited in continental-shelf seas receiving a minimal supply of terrigenous detritus from the continental interior, well supplied with nutrients from upwelling deep water off the shelf and not particularly well oxygenated.

The Senegal deposits have been exploited since 1957, and they include significant quantities of aluminium phosphate as well as the more normal tricalcium phosphate (apatite or collophane) which typifies most economic phosphate accumulations. Production in Senegal is about 1–2 million tonnes annually and reserves total well over 150 million tonnes in the known deposits. At present only some 50 000 tonnes of the phosphate output is being used for domestic fertiliser production. Phosphoric and sulphuric acid plants and fertiliser production facilities are being built. The aim is to produce nearly 250 000 tonnes of phosphoric acid per year, partly for export, over 150 000 tonnes of ammonium phosphate and over 200 000 tonnes of fertilisers. Sulphuric acid production will be based on imported sulphur. There are other potentially economic phosphate deposits in the basin, including a large one in southern Mauritania, where reserves are estimated at 30–40 million tonnes.

High-grade phosphate rock is Togo's only significant mineral resource to date, and it has been exploited since 1961. Around 2.5 to 3 million tonnes of beneficiated product grading at 36% P_2O_5 are produced annually, mostly for export to western Europe's fertiliser industry. There are plans to install plants for processing the phosphate locally, and to export the more valuable products, especially phosphoric acid.

The phosphate horizon in the Eocene sediments of the Dahomey Basin extends from Togo across Benin to southwestern Nigeria. The commercial viability of the deposits in Benin has yet to be fully evaluated, while the Nigerian phosphate in the eastern parts of the basin is at present not an economic proposition, though it has been extensively investigated.

The Iullmedden Basin also contains promising phosphate deposits, especially along the line of the Sudanese Strait and Gao Trough. In Mali, the Tilemsi valley deposits (about 150 km north of Gao) are estimated to contain about 25 million tonnes of rock averaging 27% P_2O_5. Production commenced here on a small scale in 1976, with an annual output of 2000 tonnes, and in 1981 a crushing plant with a 10 000 tonne annual capacity began operating at

Bourem (north-west of Gao, on the Niger). Output is planned eventually to exceed 200 000 tonnes annually. Further south-east along the Niger valley, the Tapoa phosphate deposit near Say in western Niger (just south of Niamey, on the Niger River), discovered in 1969, may have as much as 500 million tonnes, averaging 23% P_2O_5. On the eastern margin of the Basin, a small phosphate deposit near Tahoua is being exploited to meet domestic needs, and production in 1981 was 4000 tonnes. The Tahoua deposit has its lateral equivalents in the Sokoto sector of the basin, in Nigeria, but the lowermost Tertiary sediments here are not a promising prospect, though phosphatic nodules occur in places, and there is a phosphatic horizon in the Gamba Formation.

12.2.2 Other minerals

There are several other economically useful deposits to be found in sediments of the younger basins, but they are mostly of a type only suitable for exploitation on a local scale.

The plateau-forming oolitic and pisolitic ironstones in the Cretaceous sequence of the Bida Basin have long been recognised as potential *iron ores*. The beds reach 15 m thick in places, and the highest grades are found in the Agbaja Plateau region near Lokoja. Grades range from 30 to 60% Fe, but there are considerable amounts of phosphate (2% P_2O_5) and sulphur (0.1%) as impurity. Those make the ores unattractive at present, though the quantities are large, perhaps as much as two billion tonnes. The ores are mainly of primary sedimentary origin, upgraded to some extent by subsequent lateritic weathering. Further south, near Enugu, there are detrital and rubbly, sandy and clayey lateritic ironstones formed on beds of the Upper Coal Measures. There may be 50 million tonnes of potential ore in this area, grading over 30% Fe on average. The somewhat younger oolitic and pisolitic ironstones of the Continental Terminal in northwestern Nigeria and southern Niger have also been cited as potential iron ores. Although they cover vast areas, they are relatively thin, and mining on any scale would result in considerable environmental disruption. Similar considerations may apply to the very large deposits near Say, where an estimated 650 million tonnes of ore occur, averaging just over 50% Fe, but with a phosphate content of around 1%.

It seems unlikely that any of the iron ores in the younger sedimentary basins will be exploited in the foreseeable future. For one thing, they are nearly all characterised by significant phosphate impurity, but more important, West Africa as a whole is already very well endowed with iron ores in Precambrian rocks (Secs 3.9.2, 4.7.5, 6.5). The same may well apply to *bauxite*, which is known to occur also at Say, and there are other occurrences, for example near Agadem in the western Chad Basin. Weathering of kaolinitic clays in the Cretaceous–Tertiary sediments could well produce bauxite under favourable conditions. There are several million tonnes of *manganese* ore at Ansongo in Mali, south-west of Gao. The ore has been known for several years, but has not been exploited on any scale since the 1960s. It appears to be stratiform deposits in the Continental Terminal, and its origin may be analogous to that of the iron ores occurring among these sediments.

Sulphides of *lead* and *zinc* (galena and sphalerite), associated with smaller amounts of *copper* (chalcopyrite), have long been known from the Benue Trough. They form *en echelon* lodes and veins infilling open spaces, along a relatively narrow NE–SW belt extending roughly up the middle of the trough, from Abakaliki in the south to near Gombe in the north. They mainly occupy the axes of anticlinal folds, often occurring where these are intersected by faults, and they are characterised by numerous vughs and cavities. Trends of the individual lodes and veins are N–S or NW–SE, they are steeply dipping and they vary in length from a few metres to 2 km, in width from a few centimetres to several metres. Their vertical extent does not appear to exceed 150 m or so, and geophysical and geochemical exploration has so far failed to discover other bodies at depth.

The host rocks are mainly shales in the south, silicified limestones in the middle part of the belt, and sandstones in the north. Gangue minerals include quartz and siderite, as well as pyrite and marcasite, and cerussite and anglesite are developed in the weathered zone. In the middle Benue Trough, where the host rocks are mainly limestones, *barite* is a common associate, often forming separate veins on its own or with quartz. Here, too, *fluorite* occurs as a gangue mineral with the galena and sphalerite. These two minerals could be of local use for the oil industry (barite for drilling muds) and for the steel industry (fluorite for fluxing). In places, the galena contains up to 5 kg of *silver* per tonne, and native silver was once mined at Arufu. There is also some *cadmium* in the sphalerite, which could be won as a by-product of zinc refining.

Production of lead and zinc from these ores is sporadic and intermittent and has rarely amounted to more than a few hundred tonnes annually. The total amount of ore probably does not much exceed a million tonnes or so, and the small size of the bodies precludes large-scale mining. There is, however, some scope for exploitation to fill local needs.

The ores are undoubtedly of hydrothermal origin, and for many years it was supposed that they were associated with the igneous activity that is also widespread in the Benue Trough (Part III). However, many of the veins are far from igneous centres and fluid inclusion studies indicate temperatures of formation of less than 150°C. Moreover, isotopic studies of the lead suggest that the deposits were formed from circulating ground waters (brines) that leached metals from the basement as well as from the sediments, and deposited them in favourable sites after the sediments were folded and faulted. Present-day analogues of these brines form springs and seepages at many places and provide important local sources of *salt*.

Limestones suitable for *cement* manufacture occur in a number of places in the younger sedimentary basins, and cement industries have been established near several of them, for example, in southern Benin, in various parts of Nigeria, in southern Niger and in Senegal. An additional resource in some of the associated shales is *gypsum*, which is also used in the manufacture of cement and occurs in exploitable tonnages in southern Niger. Black and white limestones in the Abakaliki area of Nigeria have been used for ornamental chippings.

Still more abundant are *clays* suitable for the manufacture of bricks, pottery and other ceramics, and many of the clay-based industries in West Africa make use of clays from the younger basins. Kaolinitic clays are common, because of the tropical climatic conditions under which they formed, and they may be of high purity (e.g. in the Illo Formation, Sec. 10.2.1, and in the Bida Basin), with nodular and pisolitic textures that indicate some diagenetic enrichment and perhaps even incipient 'bauxitisation'. Refractory *fireclays* are associated with the Coal Measures in southern Nigeria.

Some of the sandstones have been well sorted, especially those deposited in high-energy environments in some of the coastal basin sequences, and in places they are nearly pure quartz, suitable for *glass sands*.

Sediments of the Chad Formation in the upper Benue Trough and in the Chad Basin itself have intercalations of *diatomite*, used as a filler and a filter aid, and in some insecticides. The best deposit so far found is at Bularaba, south-west of Maiduguri, containing over 50 000 tonnes of good-quality diatomite.

Bibliography

Agagu, O. K. and C. I. Adighije 1983. Tectonic and sedimentation framework of the lower Benue Trough, southwestern Nigeria. *J. Afr. Earth Sci.* 1, 267–74.

Ayoola, E. O. 1983. Hydrocarbon distribution pattern and deep prospects in the Niger Delta, Nigeria. *J. Afr. Earth Sci.* 1, 145–52.

Bigotte, G. and E. Malinas 1973. How French geologists discovered Niger uranium deposits. *World Mining* April, 34–9.

Bowden, P., J. N. Bennett, J. A. Kinnaird, J. E. Whitley, S. I. Abaa and P. K. Hadzigeorgiou-Stavrakis 1981. Uranium in the Niger–Nigerian Younger Granite province. *Mineral Mag.* 44, 379–88.

Ejedawe, J. E. 1981. Patterns of incidence of oil reserves in Niger Delta Basin. *Am. Assoc. Petrol. Geol. Bull.* 65, 1574–85.

Farrington, J. L. 1952. A preliminary description of the Nigerian lead–zinc field. *Econ. Geol.* 47, 583–608.

Ford, S. O. 1981. The economic mineral resources of the Benue Trough. *Earth Evol. Sci.* 1, 154–63.

Kogbe, C. A. and A. U. Obialo 1976. Statistics of mineral production in Nigeria (1946 to 1974) and the contribution of the mineral industry to the Nigerian economy. In *Geology of Nigeria*, C. A. Kogbe (ed.). Lagos: Elizabethan Press.

Notholt, A. J. G. 1980. Economic phosphatic sediments: mode of occurrence and stratigraphical distribution. *J. Geol. Soc. Lond.* 137, 793–805.

Offodile, M. E. 1984. The geology and tectonics of the Awe brine fields. *J. Afr. Earth Sci.* 2, 191–202.

Olade, M. A. 1976. Metallic mineral resources and exploration potential of Nigeria. In *African geology*, H. Tsegaye and A. Badejoko (eds.). Khartoum: Dept of Geology, University of Khartoum.

Petters, S. W. 1982. Petroleum geology of Benue trough and southeastern Chad basin, Nigeria. *Am. Assoc. Pet. Geol. Bull.* 66, 1141–9.

Prasad, G. 1983. A review of the early Tertiary bauxite event in South America, Africa and India. *J. Afr. Earth Sci.* 1, 305–14.

Ukpong, E. E. and M. A. Olade 1979. Geochemical surveys for lead–zinc mineralization, southern Benue Trough, Nigeria. *Inst. Min. Met. Trans. B* 88, 81–92.

Whiteman, A. J. 1983. *Nigeria: Its petroleum geology, resources and potential*. London: Graham & Trotman.

Part III

MESOZOIC TO CENOZOIC IGNEOUS ACTIVITY IN WEST AFRICA

13 *Introduction to anorogenic magmatism*

SUMMARY

Anorogenic magmatism in West Africa began in the Palaeozoic, soon after the Pan African event, but most of it occurred from Mesozoic times onwards. A number of petrographic provinces can be identified.

Permo-Triassic dolerites are widespread in the southwestern part of the region, emplaced in fracture systems that developed during the separation of Africa from North and South America. The Younger Granites of Nigeria and Niger form a discontinuous north–south belt nearly 1500 km long, with ages ranging from Ordovician in the north (Aïr) to Jurassic in the south (Jos Plateau). Another line of mainly Tertiary granites extends NE for about 1000 km along the Cameroun volcanic line. Carbonatites and related rocks of probable Permian age have recently been discovered in NE Mali.

There is good spatial separation between these two provinces, the Permo-Triassic intrusives being confined to the West African craton, the Younger Granites to Pan African terranes. Post-Jurassic magmatism occurred mainly in the Pan African terrane east of the craton.

Kimberlites and peralkaline intrusions were emplaced at several localities on the West African craton, some probably of Cretaceous age, others perhaps older. Otherwise, igneous activity in the Cretaceous was confined to the Benue Trough. Cenozoic alkali basaltic volcanism was most voluminous along the Cameroun line, but occurred in many other places.

13.1 The distribution of anorogenic magmatism

Igneous activity accompanying the major thermo-tectonic events that built the West African segment of the African shield was described in Part I. After the Pan African event the whole continent was effectively 'cratonised' and it has been subjected only to faulting, stretching, warping, and differential uplift. Magmatism in Africa over the last 450 Ma or so has therefore been almost entirely of **anorogenic** type.

In West Africa a number of quite well defined **petrographic provinces** (igneous provinces) can be identified on the basis of petrological and geochemical affinities, each fairly well circumscribed in space and time. Each province is related to a particular geotectonic environment, but the nature of that environment cannot always be fully evaluated, partly because the different provinces overlap in space and time. Thus, there was major igneous activity in the Mesozoic (Permo-Triassic to Jurassic) and at least some of this activity can be plausibly related to the stresses that accompanied continental separation of Africa from North and South America.

It is less easy to infer the geotectonic environment responsible for initiating anorogenic granite magmatism in Niger during the Palaeozoic and maintaining it through the Mesozoic and Tertiary in Nigeria and Cameroun. Nor is it immediately obvious why widespread basaltic volcanism occurred during the later Cenozoic. However, each of the igneous provinces can be recognised easily enough on the basis of a number of distinguishing characteristics.

Figure 13.1 summarises the igneous provinces that have been identified, separating them into two quite well-defined groups, despite the overlaps in space and time.

Palaeozoic to Jurassic This age range covers most of the igneous rocks in the older provinces, with some overlap into the Tertiary (Fig. 13.1a).

(a) Permo-Triassic dolerite sills and dykes and other intrusions occur over much of southern Mali, Liberia, Guinea, Sierra Leone and Senegal. They are confined to cratonic basement, even though the sills occur among Palaeozoic and older sediments that overlie the craton. Geochemically they have **tholeiitic**

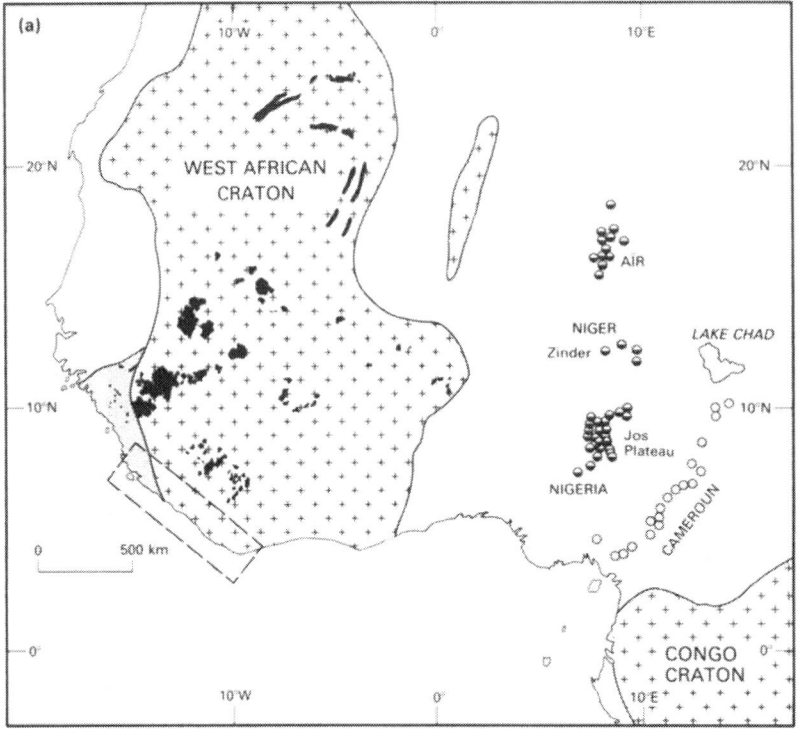

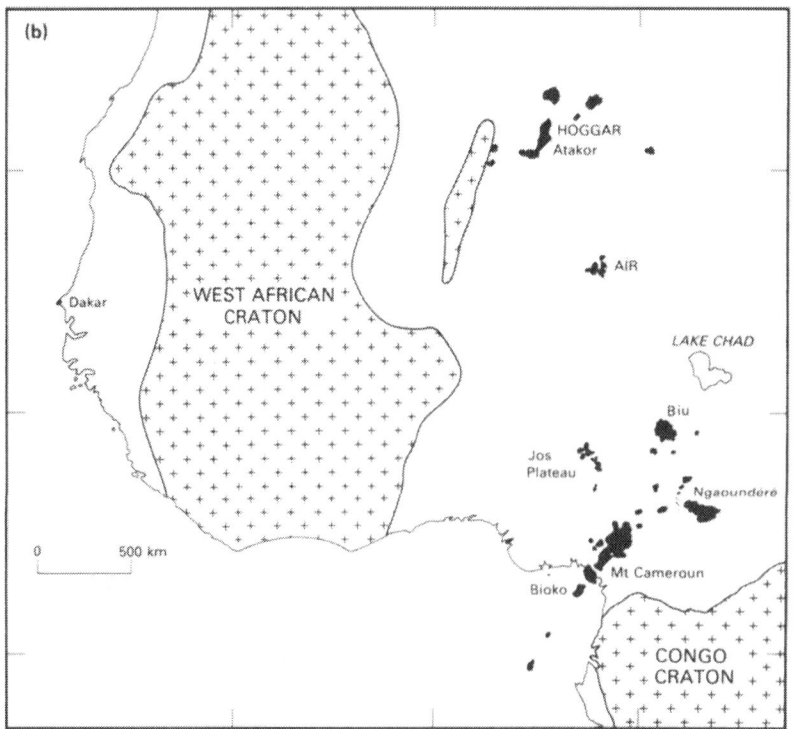

Figure 13.1 The distribution of anorogenic igneous rocks in West Africa. (a) Permo-Triassic dolerites (black) are confined to the West African craton, including the area affected by Pan African orogenesis (shaded). The NW–SE trending rectangle marks the approximate area of coast-parallel dyke swarms in Sierra Leone and Liberia. Further north, in western Mali and southern Mauritania, dykes have a generally NE–SW trend. Palaeozoic to Jurassic Younger Granites (half-filled circles) occur in three main areas along a north–south belt east of the craton. Tertiary Younger Granites of Cameroun (open circles) form a NE–SW trending belt along the Cameroun volcanic line (see (b)). The belt of Permian carbonatites and related rocks is shown as parallel broken lines. (b) Cenozoic basaltic volcanism (black) is mainly confined to Pan African areas east of the craton, and there is some overlap with Younger Granite areas, especially along the Cameroun line. Cretaceous(?) kimberlite–carbonatite occurrences are not shown, nor are the Cretaceous volcanics of the Benue Trough.

affinities. The distribution and age of these intrusions leave little doubt that their emplacement was closely related to opening of the North Atlantic ocean.

(b) Clusters and lines of **ring complexes** dominated by granitic rocks occur in four well defined areas, in each of which they encompass a different age span: Aïr (Lower Palaeozoic), Zinder (Upper Palaeozoic), Jos Plateau (Triassic to Jurassic), Cameroun (Tertiary). There is thus a southward progression of ages, and intrusions in the first three of these areas lie along a north–south belt nearly 1500 km long that coincides roughly with the northward projection of the western coastline of southern Africa. Intrusions in the fourth area lie along the NE–SW trend of the Cameroun volcanic line. All the ring complexes were emplaced into Pan African basement, and they occupy basement 'highs' which border Mesozoic sedimentary basins. The rocks are richly endowed with tin mineralisation in Nigeria (Sec. 15.5), where they are collectively called **Younger Granites** to distinguish them from the Older Granites of the Pan African.

(c) A strongly undersaturated Permian alkaline province of ring complexes dominated by nepheline syenite with **carbonatite** was discovered in the early 1980s in north-east Mali, along the eastern margin of the West African craton.

Cretaceous to Cenozoic and Recent   Igneous rocks in the lower part of this age range (i.e. items (a) and (b) below) are volumetrically too small to show on Figure 13.1b.

(a) **Kimberlites** and **peralkaline** rocks and carbonatites are confined to the West African craton. Radiometric age determinations range from Proterozoic to Cretaceous, but the older ages may not represent the time of emplacement. Some of the kimberlites are diamondiferous (Ch. 16).

(b) Mainly basaltic magmatism occurred in the Benue Trough during the Upper Cretaceous, where the rift faulting presumably provided pathways for the rise of magmas. Igneous activity accompanied both sedimentation and subsequent folding, and there is evidence that the rocks were altered as a result of emplacement into wet sediments.

(c) In the Cenozoic there were major effusions of lava along the Cameroun volcanic line and from the Tibesti volcano. There was also significant volcanism in the Hoggar, Aïr, the Benue Trough and the Jos Plateau. Cenozoic volcanic rocks in the last two of these are sometimes grouped with the Cameroun volcanics into a Gulf of Guinea province, which includes the offshore island chain from Bioko (formerly Fernando Po) to Pagalu (formerly Annobon). There was also minor Cenozoic volcanism in the Cape Verde peninsula in Senegal, possibly related to evolution of the Cape Verde Islands. Throughout the Cenozoic volcanic province as a whole, the rocks are dominated by **alkali basalts**, with subordinate phonolites, trachytes and rhyolites. Most of the volcanism took place in the Mio-Pliocene to Quaternary interval, and locally it has continued up to the present (e.g. Mount Cameroun). As in the case of the anorogenic granites, much of the volcanism broke out on basement 'highs' bordering sedimentary basins.

Bibliography

Black, R. and M. Girod 1970. Late Palaeozoic to Recent igneous activity in West Africa and its relationship to basement structure. In *African magmatism and tectonics*, T. N. Clifford and I. G. Gass (eds). Edinburgh: Oliver & Boyd.

Bowden, P., O. van Breemen, J. Hutchinson and D. C. Turner 1976. Palaeozoic and Mesozoic age trends for some ring complexes in Niger and Nigeria. *Nature* **259**, 297–9.

Liégeois, J. P., H. Bertrand, R. Black, R. Caby and J. Fabre 1983. Permian alkaline undersaturated and carbonatite province, and rifting along the West African Craton. *Nature* **305**, 42–3.

Williams, H. R. and R. A. Williams 1977. Kimberlites and plate-tectonics in West Africa. *Nature* **270**, 507–8.

Wright, J. B. 1976. Volcanic rocks in Nigeria. In *Geology of Nigeria*, C. A. Kogbe (ed.). Lagos: Elizabethan Press.

14 *The Permo-Triassic dolerites and carbonatites*

SUMMARY

The Permo-Triassic dolerites intrude rocks of the West African craton and the overlying Palaeozoic sediments. They are dominated by dykes and sills. The Freetown igneous complex is a layered gabbro–anorthosite intrusion, in which the layering dips concentrically inwards to define a cone-shaped structure. All the rocks in this province fall in the age range 275–175 Ma, and they are of tholeiitic affinities. The distribution of the West African dolerite intrusions, when considered with those of similar type and age in other countries bordering the Atlantic, suggests that they were emplaced into a strongly tensional environment, during the early stages of formation of the Atlantic Ocean.

A tensional environment also controlled emplacement of the newly discovered nepheline syenite–carbonatite complexes on the eastern margin of the West African craton, which are dated as Lower Permian (c. 265 Ma) and were emplaced in a well defined continental rift setting.

14.1 The dolerites

Figure 13.1a shows the distribution of the numerous Mesozoic dolerite intrusions in West Africa. They occur mainly as dykes and sills, intruding rocks of the West African craton and the Palaeozoic sediments that overlie them. Some of the sills in the southern part of West Africa are also shown in Figure 9.1. They mostly occur along a broad belt extending northeastwards from the southern end of the Bové Basin, but they are also found outside this zone. The intrusions cut rocks up to and including the Continental Intercalaire. Their upper age limit as defined by stratigraphic relationships is not younger than Lower Jurassic. Many of the dykes that occupy N–S faults and fractures in the West African craton (Secs 3.5 & 4.1) probably belong to this igneous province.

Individual dykes can be as much as 200 m across, though they are mostly much narrower than this. Dykes form swarms parallel to the coast in Sierra Leone, Liberia and Ivory Coast, and become more abundant towards the coast. Away from the coast, trends are more variable. The dykes may bifurcate and display *en echelon* patterns of emplacement, and marginal cataclasis is often found, consistent with emplacement along faults that have subsequently been reactivated. Geophysical investigations in Liberia show a predominance of NW–SE trending dykes both along the coastal belt and extending well out beneath the continental shelf. The amount of crustal dilation necessary to accommodate the total thickness of dykes that have been emplaced has been variously estimated at between 5 and 20%.

Many of the sills are several hundred metres thick and can be traced laterally for distances of 50 km or more. They are found at altitudes ranging from nearly 1500 m to below sea level and they can form minor sill 'swarms', with as many as ten sills being found within a sedimentary sequence in some places. In the southwestern Taoudeni Basin and the Bové Basin, the sills tend to form a 'step-down' pattern towards the south-west. Because of their resistance to erosion, they also form the summits of hills, e.g. Bintumani and Saionia Scarp in Sierra Leone, and Mount Linsan in Guinea.

Sills are rare east of Sierra Leone, because there are few onshore outcrops of sediments of suitable age. The **Monrovia Diabase** is a dolerite sill intruding the Paynesville Sandstone of Liberia (Sec. 9.5), and boreholes have intersected a dolerite layer about 70 m thick in the Keta Basin of Ghana, at a depth of about 3500 m. This is interpreted as an early Jurassic sill, intruded into the top of Devonian strata that are overlain by Lower Cretaceous sediments. A similar sill has been found in the Tano Basin further west.

14.1.1 Petrology and age

The rocks generally display remarkably uniform textural, mineralogical and chemical features throughout the area in which they occur. They all consist essentially of zoned labradoritic plagioclase which is sometimes phenocrystic, and ophitic augite. Olivine, orthopyroxene and low-calcium clinopyroxene (pigeonite) are common among the major constituents, and interstitial graphic quartz–feldspar intergrowths are a frequent late-stage accessory phase. Alteration of the dolerites is common, particularly in the smaller intrusions, where relict olivines occur and pyroxene has been replaced by amphibole and biotite.

Some sills display igneous banding with mineralogical variations that are interpreted as resulting from crystal settling during gravity differentiation. The Mount Kakoulima sill in Guinea, one of the largest, has dunite and wehrlite layers at the base, passing up to some 800 m of gabbros. Dykes and plugs of granophyre are associated with the dolerites and the whole assemblage has tholeiitic affinities.

Radiometric age determinations (mainly K/Ar) on samples from several localities provide an age range of 275–175 Ma, suggesting a long period of intermittent igneous activity from Permo-Triassic to lowermost Jurassic. It is possible that the coastal dyke swarms were mostly emplaced towards the end of the period, though two dykes in Sierra Leone gave ages as far apart as 230 and 185 Ma. Dykes and sills so far dated in Liberia fall in the range 192–173 Ma. The Monrovia Diabase yielded an age of 184 ± 8 Ma, whereas the Keta Basin sill gave 167 ± 5 Ma.

14.1.2 The Freetown complex

Among the largest and most interesting intrusions of this province is the layered gabbro **Freetown complex** in Sierra Leone, emplaced in gneisses and schists of the Kasila Group (Sec. 3.3.1). Its outcrop limits define an arc concave to the south-west and the layering dips radially inwards, suggesting a cone-shaped or funnel-shaped intrusion. Repeated layers include dunite, troctolite and anorthosite, as well as gabbro and olivine gabbro, and the total thickness is 6 km, made up of several distinct cycles. Dips in the layering increase from 10–20° near the margin to 45–50° near the middle, and igneous sedimentary features consistent with crystal settling are observed, such as slumping, cross bedding and mineral lamination. However, it is likely that flow differentiation and volatile-dependent crystallisation processes predominated. A rhythmic pattern of crystallisation of different combinations of mineral phases, building up from the bottom and inwards from the sides, resulted in the formation of the layering. The complex is reliably dated at close to 195 Ma.

14.1.3 The tectonic setting

In the southwestern corner of the Taoudeni Basin, notably in Sierra Leone and Guinea, the sills emplaced into Palaeozoic and Infracambrian sediments occur at progressively lower levels towards the coast. The Freetown complex is at sea level, for example, and so is the Monrovia Diabase further east. In addition, there is a marked increase in the frequency of dykes towards the coast of Sierra Leone and Liberia, and evidence from aeromagnetic surveys suggests that dykes increase in abundance on the continental shelf off Liberia. Both of these tendencies can be related to stretching and subsidence of continental crust during the early stages of disruption of Gondwanaland, when Africa separated from North and South America.

The continental margins have continued to subside since the Jurassic, to accommodate the great thicknesses of Mesozoic sediments in the coastal basins (Part II). Dolerite sills found at depths of a few kilometres in those basins (e.g. the Keta Basin) were originally intruded into sediments at or near sea level and have subsided since the time of their emplacement.

The Permo-Triassic dolerites occur almost entirely in areas underlain by the West African craton. Only in the extreme west, in the southern part of the Mauritanide belt, do they extend into strata overlying basement affected by the Pan African. In contrast, the Palaeozoic to Tertiary granites of Niger, Nigeria and Cameroun occur exclusively in Pan African terranes.

14.2 The undersaturated complexes

The undersaturated alkaline ring complexes and minor intrusions of north-east Mali, discovered in the early 1980s, occur along a roughly NNE–SSW trending line that may be 50–100 km long or more, at the northern end of the Tilemsi trough (part of the Sudanese Strait, Sec. 10.2).

The rocks comprise nepheline syenites, melteigites and ijolites, along with pyroxenites, phonolites, tinguaites and microsyenites, as well as carbonatites. They form not only ring complexes, up to 12 km across, but also smaller plugs and dyke swarms. Radiometric ages so far obtained lie between 260 and 270 Ma, consistent with emplacement in Lower Permian times.

The intrusions are largely masked by Cretaceous–Tertiary sediments and occur within a rifted depression that also contains over 2000 m of older continental sandstones and conglomerates. These are of indeterminate but probably Lower Palaeozoic age and probably derived from erosion of Proterozoic sediments of the Taoudeni basin to the west.

The linear zone west of the Adrar des Foras, extending both northwards and southwards along the West African craton margin (marked by positive gravity anomalies, Sec. 6.3.6), may therefore have been characterised by intermittent rifting from shortly after the end of the Pan African event until early Tertiary times.

The Permian age of these complexes is close to that of the dolerite intrusions on the craton. It seems plausible to relate the emplacement of both sets of igneous rocks to stresses that were felt throughout the world in the early stages of fragmentation of supercontinents, one consequence of which was formation of the Atlantic ocean.

14.3 Economic potential of the Permo-Triassic intrusions

Dolerites Few minerals of economic importance are associated with minor basic intrusions, and the West African dolerites are no exception to this rule. There is always the possibility of contact metasomatism or hydrothermal mineralisation, but the *titanium*-rich magnetite and *copper* showings recorded near Nioro in the southern Taoudeni Basin, where dolerite intrudes limestone, are almost certainly of curiosity value only. Limited amounts of *iron ore* have formed by lateritic alteration of larger sills in Guinea. There are Ti-rich magnetites in the Freetown complex, and sulphides of *copper* and *nickel* are associated with *platinum* and *gold* in these rocks; but these too are only of mineralogical interest, as the grades are very low and the total amount involved cannot be large, for the intrusion itself is relatively small.

The most useful role for the dolerites in economic terms is as *crushed rock* for aggregate, roadstone or rail ballast. Dolerites emplaced as sills in relatively soft sediments are often the hardest and most suitable rocks for constructional purposes to be found in an area, and they often have the additional advantage of being fairly uniform, so that they break into more or less equidimensional fragments.

Carbonatites and related rocks *Uranium* mineralisation reported in northeastern Mali may be related to these rocks, which may also carry *thorium* and *rare earth elements*. Carbonates elsewhere are rich in *phosphate*, others may contain ores of metals as geochemically diverse as *copper* and *niobium*. Minerals such as these could also be present in the Mali rocks.

Bibliography

Akpati, B. N. 1978. Geologic structure and evolution of the Keta Basin, Ghana, West Africa. *Geol. Soc. Am. Bull.* **89**, 124–32.

Beckinsale, R. D., J. F. W. Bowles, R. J. Pankhurst and M. K. Wells 1977. Rubidium–strontium age studies and geochemistry of acid veins in the Freetown complex, Sierra Leone. *Mineral Mag.* **41**, 501–12.

Behrendt, J. S. and C. S. Wotorson 1970. Aeromagnetic and gravity investigations of the coastal area and continental shelf of Liberia, West Africa, and their relation to continental drift. *Geol. Soc. Am. Bull.* **81**, 3563–74.

Behrendt, J. S., J. Schlee, J. M. Robbo and M. K. Silverstein 1974. Structure of the continental margin of Liberia, West Africa. *Geol. Soc. Am. Bull.* **85**, 1143–58.

Black, R. and M. Girod 1970. Late Palaeozoic to Recent igneous activity in West Africa and its relationship to basement structure. In *African magmatism and tectonics*, T. N. Clifford and I. G. Gass (eds). Edinburgh: Oliver & Boyd.

Culver, S. J. and H. R. Williams 1979. Late Precambrian and Phanerozoic geology of Sierra Leone. *J. Geol. Soc. Lond.* **136**, 605–18.

Fabre, J. 1976. *Introduction à la Géologie du Sahara Algérien*. Algiers: SNED.

Furon, R. 1963. *Geology of Africa* (English edn). London: Oliver & Boyd.

Liégeois, J. P., H. Bertrand, R. Black, R. Caly and J. Fabre 1983. Permian alkaline undersaturated and carbonatite province, and rifting along the West African craton. *Nature* **305**, 42–3.

May, P. R. 1971. Patterns of Triassic diabase dikes around the North Atlantic in the context of predrift positions of the continents. *Geol. Soc. Am. Bull.* **82**, 1285–92.

Wells, M. K. and J. F. W. Bowles 1981. The textures and genesis of metamorphic pyroxene in the Freetown Intrusion. *Mineral Mag.* **44**, 245–56.

15 *The Younger Granites*

SUMMARY

The Younger Granite ring complexes of West Africa extend from Aïr to Cameroun and range in age from Palaeozoic to Tertiary. They are dominated by granites and are emplaced in Pan African basement rocks, on 'swells' bordering sedimentary basins.

The complexes probably represent the root zones of ancient volcanoes of the type characterised by caldera collapse. Early stages of magmatic activity involved the eruption of large volumes of rhyolitic ignimbrites. Further eruptions occurred along more or less circular ring faults. Magma subsequently solidified in these fractures and formed marginal ring dykes of granite porphyry that define the outer limits of some complexes. Inside the peripheral ring fracture, a variety of mainly granitic rocks was emplaced, both as massive ring dykes and as more or less cylindrical stocks and bosses. The ideal pattern is one of concentric intrusions, becoming progressively younger towards the centre, but many complexes depart from this ideal.

The dominant granites of the complexes range from peraluminous to peralkaline in composition and they are associated with smaller amounts of syenite, gabbro and anorthosite. Some of these magmas originated in the upper mantle, but the overwhelming preponderance of granite suggests that there was a contribution from crustal melting also.

Regional tectonic controls on emplacement of the complexes remain a matter of speculation. On the local scale, emplacement of individual complexes must have been controlled by basement fracture systems, but these are not easily identified.

The Younger Granites of Nigeria in particular are famous for their tin (cassiterite) mineralisation, which is mainly associated with the biotite granites. These rocks also contain significant quantities of the niobium-rich mineral columbite as an accessory. Most of the workable deposits of cassiterite and columbite are in alluvial concentrations. The peralkaline granites also contain accessory uranium-bearing minerals, which probably provided the primary source for the sedimentary uranium deposits of Niger.

15.1 Introduction

The numerous ring complexes collectively termed the Younger Granites are now found along a nearly 1500 km long belt in areas of Pan African basement uplift surrounded by sedimentary basins that achieved their main development in the Cretaceous (cf. Fig. 13.1a). The Younger Granites have been studied in most detail in Nigeria, partly for their intrinsic interest, providing comparative data for study of similar formations elsewhere in the world, but mainly because in the early 1900s they were recognised as the source of rich alluvial cassiterite deposits that had long been known to exist on and around the Jos Plateau (Sec. 15.5).

Detailed field mapping of the ring complexes has demonstrated a consistent succession of magmatic activity from volcanism to plutonism associated with the emplacement of mainly granitic melts at high levels in the crust. The most striking petrographic feature of the whole province is the overwhelmingly acid nature of the rocks and the uniformity of rock types found in all areas. Over 95% of the rocks can be classified as rhyolites, quartz–syenites or granites,

with basic rocks forming the remaining 5%. Many of the rocks have strongly alkaline to peralkaline compositions, others are aluminous to **peraluminous**.

Initial stages in development of the complexes involved extrusion of vast amounts of acid lavas, tuffs and ignimbrites, now only partly preserved as a result of subsidence along ring faults. Almost everywhere these rhyolitic rocks directly overlie the metamorphic basement, which means that the Younger Granites were emplaced in uplifted areas that were undergoing erosion. Granitic **ring dykes** are the major component of most complexes, ranging from 5 km or less to over 30 km in diameter, and varying in plan from the polygonal to circular or crescentic, and through more irregular shapes to simple stocks and bosses. Some complexes have a broadly concentric pattern, indicating that the activity was confined to one area, but others have overlapping rings, because the centre of activity migrated with time.

In most Younger Granite complexes, both volcanic and intrusive rocks are confined to roughly circular areas bounded by ring dykes or ring faults. The volcanic rocks were downfaulted in great collapse structures that formed calderas at the surface. The volcanics are fairly well stratified and they are tilted to varying degrees, as a result partly of sagging and fault drag accompanying caldera collapse and partly of emplacement of later granites. Although the volcanics probably once extended well beyond the present limits of the complexes, they are now only preserved in these downfaulted regions. Virtually all traces of their former extent outside the caldera walls have long since been eroded away. Even where ring complexes have no associated volcanics at all, this is probably because they have been removed by erosion rather than because they were never erupted. The ring dykes were probably emplaced by mechanisms involving underground **cauldron subsidence**, though as the intrusive forms are quite varied other processes of intrusion must also have been involved. An important feature of several complexes is the occurrence of **screens** of crystalline basement rocks between successive granitic intrusions.

15.2 Evolution of the complexes

Figure 15.1 shows the likely sequence of evolution of the Younger Granite complexes, in highly simplified and abbreviated form. Eruption of the volcanic components probably followed a pattern broadly similar to that found in modern calderas, and two main stages can be recognised.

15.2.1 Eruption of early rhyolites
Where significant volumes of volcanics are preserved, the succession generally commences with layered sequences of rhyolites, **ignimbrites** and welded tuffs, bedded air-fall tuffs, agglomerates and occasional basalts and trachytes.

Most of the early rhyolites are typical ignimbrites or welded tuffs, with original fragmental textures obscured by welding and later devitrification and recrystallisation. Many lack any sign of fragmental textures, and may be lavas. Compositionally, the early rhyolites consist of quartz and alkali feldspar phenocrysts in a fine-grained devitrified ground mass.

Little is known of the pre-caldera structures formed by these early effusions, but the dominance of ignimbrites suggests that they may have been plateaux or very flat shield volcanoes. The surface activity must have been preceded by the formation of a magma chamber high in the crust (Fig. 15.1).

This caused crustal doming and so promoted development of the ring fracture that controlled surface caldera collapse. Evidence for pre-caldera doming is found in the **cone sheet** swarms round a few complexes, and circular lineaments visible on satellite photographs round others. The cone sheets are normally thin curved intrusions of felsite or granophyre, dipping inwards at angles of between 30° and 60° and projecting to a focus some 5 km below the present surface.

15.2.2 Eruption of later rhyolites
Massive porphyritic rhyolites up to several hundred metres thick and xenolithic in places, overlie the bedded volcanics of the early rhyolites in several complexes. They probably represent caldera-filling eruptions and appear to occur in two main forms: (a) crystal-rich ignimbrites or welded tuffs, often banked up against caldera walls and erupted as ash flows from the ring fracture which controlled the surface caldera collapse and underground cauldron subsidence (Fig. 15.1), and (b) massive bodies representing original highly viscous magma extruded as thick flows and dome-like masses (**tholoids**). The later rhyolites are all crystal-rich, with abundant quartz and alkali feldspar phenocrysts.

It is possible that the main phase of caldera subsidence was initiated during this eruptive episode. The high phenocryst content of the later rhyolites indicates that the magma had reached an advanced stage of crystallisation within the subvolcanic reservoir. This probably contributed to the build-up of gas pressures that initiated the large ignimbrite eruptions. The resultant partial evacuation of the magma chamber started the process of collapse and the formation of a surface caldera. The ring fracture which now formed the main pathway for eruption of the magma, probably as a **fluidised system**, was widened by the violent nature of the eruptions. It subsequently became filled with the relatively gaspoor and more viscous ring dyke porphyries described in the next section. The amount of caldera subsidence can be estimated from the thickness of volcanics preserved in Younger Granite complexes, and it was of the order of 500 to 1000 m.

Post-caldera eruptions are known in some modern calderas, but either these did not occur in the West African complexes or their products have since been lost through erosion. In any case the igneous activity often shifted sideways, so that renewal of volcanism on sites of previous collapse may have been relatively uncommon. Clastic sediments are preserved in a few

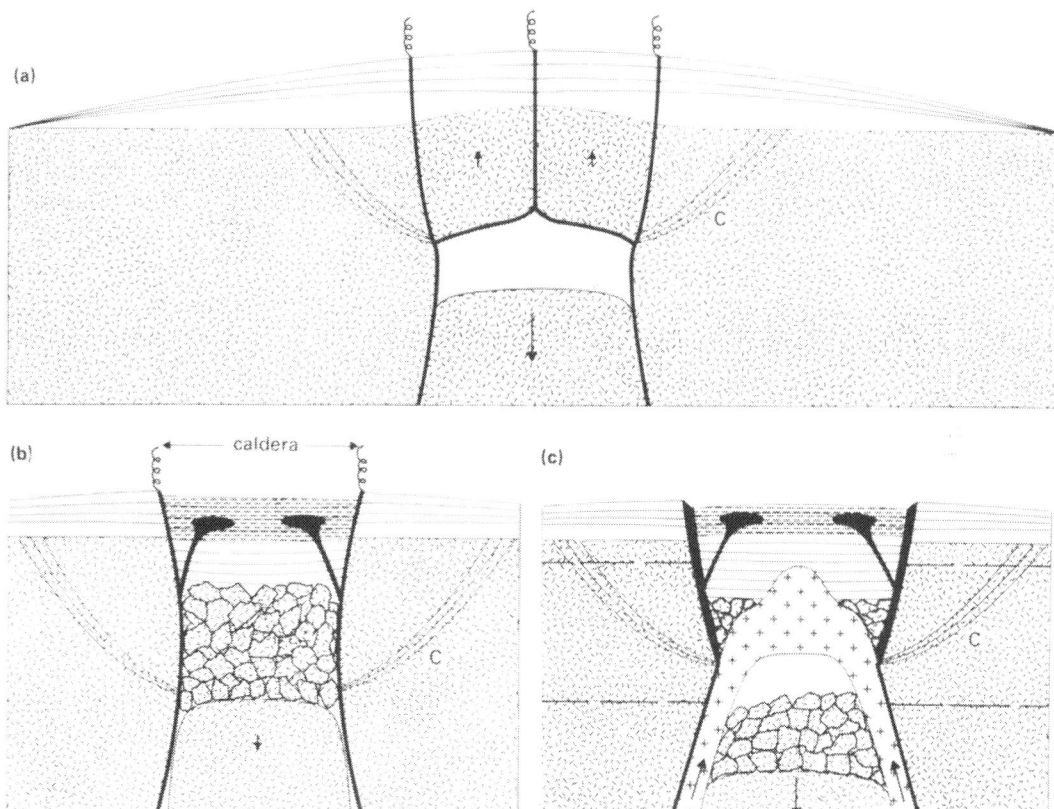

Figure 15.1 Highly schematic cross sections to show stages in the development of Younger Granite complexes in simplified and abbreviated form. (a) A mass of granite (shaded) rises high into the crust (basement gneisses, migmatites and granites), supplied along ring fractures from below. Emplacement is accompanied by doming or swelling of the overlying crust and by initial subsidence of the underlying crustal block. Cone sheets (C) may be emplaced. Predominantly rhyolitic magmas are erupted, mainly as ignimbrites, from arcuate vents along the ring fracture and also from central vents. (b) Rapid eruption of ignimbrites empties the magma chamber so that the overlying crustal cylinder breaks up and collapses into the resulting void, and there may also be further subsidence of the crustal block beneath. Early volcanics are downfaulted into the surface caldera, which is filled by eruptions of later rhyolites in the form of both ignimbrites (dashes) and viscous lava masses (black). These are mainly supplied by magma rising along the ring fractures. Caldera subsidence continues during eruption of the rhyolites. (c) Eruptions cease and the ring fractures are filled with granite porphyries to form marginal ring dykes (black). Emplacement of granites (crosses) occurs by subterranean cauldron subsidence: intermittent large-scale sinking of the roughly cylindrical crustal block beneath the caldera is accompanied by the rise of further supplies of granitic magma from depth. The granites are emplaced into the upper crust, intruding and sometimes doming the earlier volcanics. The heavy broken lines indicate approximate upper and lower limits of present-day erosion levels seen in complexes in different places.

complexes, evidently deposited in caldera lakes. They consist of sandstones and siltstones and the maximum recorded thickness is 300 m. These sediments indicate that substantial intervals of time could elapse between different phases of activity in the complexes.

15.2.3 Marginal ring dykes

The outer limits of some Younger Granite complexes are defined by a ring dyke of granite porphyry (Fig. 15.1c), though in places they occur separate from the main part of the complex. Most of these ring dykes are incomplete, but sometimes there is evidence of a continuous ring fault linking separate segments. Contacts with the basement rocks are usually steep and volcanic rocks are typically exposed within the ring dykes, not outside them. The characteristic rock type of the marginal ring dykes is a distinctive granite porphyry which is greenish in fresh hand specimens, and is notable for the presence of the iron olivine fayalite.

15.2.4 Internal ring intrusions

The main central part of each complex is occupied by granite intrusions. They vary greatly in number and structural complexity, with outcrop patterns ranging from circular to crescentic to irregular. Some are ring dykes, but they differ from the marginal ring dykes in being much thicker and the majority are generally composed of granite rather than porphyry. They often have outward dipping contacts and many intrusions show flat-lying or gently dipping roof sections, giving them the form of a domed sheet – (Fig. 15.1c). With increasing thickness of the roof section, these grade into massive cylindrical bodies of granite – stocks and bosses.

There is a more or less concentric pattern to these later granitic bodies, successive intrusions having smaller diameters but retaining the shape of earlier members. Each intrusion was emplaced along the inner margin of its predecessor as cauldron subsidence continued, though the coincidence was often only approximate.

Where migration of centres occurred, each centre must have gone through a complete ring complex cycle before the focus of activity shifted, for ring dykes of later centres cut across those of preceding ones.

The internal granites include the whole range of Younger Granite compositions. They are composed of perthitic alkali feldspar, quartz and usually less than 10% of various mafic minerals, which can be used to name and classify the granites. Although there is a certain amount of overlap and gradation between the different types, three broad groups can be identified. They are listed below in the approximate order of their emplacement in individual complexes, as worked out from their field and contact relationships.

(a) Granites and porphyries occur with fayalite, hedenbergite and hastingsite or arfvedsonite. These closely resemble the porphyries of the peripheral ring dykes and are characteristically greenish in hand specimen, and are generally among the earliest of the internal intrusions.

Accessory minerals include ilmenite, zircon, fluorite and fergusonite. A subgroup of coarse hastingsite–biotite granites is transitional towards the biotite granites.

(b) Aegirine–arfvedsonite and riebeckite granites are generally equigranular in texture, though some may be porphyritic. These are peralkaline rocks and some small intrusions may contain abundant prismatic albite, with accessory pyrochlore, cryolite and amblygonite.

(c) Biotite granites are granular rocks with a wide range of grain sizes and they are the principal repositories of the mineralisation in the province, especially the medium- to fine-grained varieties. They were almost invariably emplaced last. Biotite forms only about 5% or less of these rocks, which tend to be peraluminous. Common accessories are fluorite and iron oxides, and there is evidence of late introduction of secondary albite. Veins of quartz–mica–topaz greisen contain cassiterite, and the albitic biotite granites contain accessory columbite.

The generally later biotite granite intrusions are believed to represent large volumes of new and relatively hydrous magmas that did not rise to such high levels as the somewhat drier pyroxene and amphibole granite melts which usually preceded them. Whereas the early intrusions are generally arcuate or crescentic, with a close spatial relation to the outer ring dykes, the biotite granites are more often discordant with them. They tend to have irregular shapes and shallow outward dipping contacts and they sometimes cut across and partly obliterate earlier ring structures. Being relatively hydrous, the magma that formed the biotite granites was richest in residual constituents, notably tin and niobium; hence their economic importance.

The Younger Granites are virtually all one-feldspar hypersolvus granites containing perthitic orthoclase or microcline (where albite occurs, it is often of secondary origin). In this respect they are readily distinguished from Older Granites of the basement, which are invariably two-feldspar subsolvus granites containing microcline and sodic to intermediate plagioclase.

Among the minor rock types, the syenites show gradations towards the fayalite and arfvedsonite granites through increasing quartz content. The gabbros contain labradorite, augite and brown hornblende and sometimes olivine. They may be associated with hornblende gabbros and diorites in net-veined masses formed by later granite intrusions. Anorthosites are virtually confined to the Aïr sector of the province. They are generally coarse-grained, consisting of andesine–labradorite, augite, amphibole and biotite. Where they occur, all these minor types were generally emplaced early in the intrusion sequence.

15.2.5 External dykes

As well as concentric cone sheet swarms of felsite or granophyre occurring outside some complexes, occasional radial dykes of felsite are found. Dykes of dolerite and basalt – mostly rather altered – are widely distributed in and around the ring complexes, at least in Nigeria; but their total volume is very small. They have two principal trends, NNE and NNW–NW. Most are probably of Younger Granite age, but some may belong to the late Pan African basic dyke suite mentioned at the end of Section 6.3.3.

15.3 Regional distribution and setting

The north–south belt occupied by Younger Granite complexes (Fig. 13.1a) is a discontinuous one, and must not be regarded as a single crustal arch. It must rather be seen as a series of crustal domes beneath which magmas were generated at different times. The complexes in the Aïr appear to have been emplaced in the Lower Palaeozoic some 430–380 Ma ago, not long after the Pan African event. In southern Niger, the complexes are somewhat younger, of Carboniferous age (c. 340–290 Ma), while the Nigerian complexes have Mesozoic ages ranging from 190 to 145 Ma, placing them mostly in the late Triassic to Jurassic interval, and thus overlapping with the basic intrusive activity further west, on the craton (Ch. 14). Granites of the elongate NE–SW Cameroun belt have yielded Tertiary ages, in the 65–35 Ma range. They narrowly pre-date the onset of Cenozoic volcanism along the Cameroun line (Ch. 16).

The various rock types of the Younger Granites are not evenly distributed throughout the province. Gabbroic rocks (including anorthosite) and syenites are more abundant in the Palaeozoic Aïr sector in the north and the Cenozoic Cameroun belt in the south than in the Mesozoic complexes of Nigeria and southern Niger; and peralkaline granites are least abundant in Niger. Conversely, biotite granites are much less well represented in the northern and

southern regions than in Nigeria, which probably accounts for the relatively poor mineralisation prospects of these regions (Sec. 15.5).

The early volcanics are also unevenly distributed, but this is probably a function more of erosion than of magma type. They are best preserved in southern Niger and in the northern parts of Nigeria, where the biotite granites are exposed only in their roof zones. Further south, in central Nigeria, volcanics are found only as small downfaulted remnants among the granites; south of the Jos Plateau they are absent, and massive intrusions of biotite and hastingsite–biotite granites are common. This distribution pattern is attributed to uplift along the flanks of the Benue Trough which enabled erosion to cut much deeper into the southern ring complexes than the northern ones, where post-magmatic uplift was less.

Still further south, the generally smaller Tertiary complexes of the Cameroun line have also lost most of their volcanic rocks through erosion, presumably as a result of the considerable uplift that occurred along this line in association with later Cenozoic activity. Volcanics have also largely disappeared from the Aïr complexes in the north, which are much older and have therefore been subjected to more prolonged erosion.

The north–south alignment of basement swells occupied by the Palaeozoic–Mesozoic Younger Granites of Niger and Nigeria lies on the northward projection of the coastline south of the Gulf of Guinea. This was the line of continental separation between the southern halves of Africa and South America in the Cretaceous. The NE-trending Cameroun line of Tertiary Younger Granites is parallel with the trend of the Benue Trough, which developed as the continents separated.

In southern Niger and Nigeria, the complexes lie a short distance east of the zone of Pan African metasedimentary belts, where the basement consists of gneisses and migmatites and Older Granites. The main concentration of complexes on the Jos Plateau (Fig. 13.1a) lies virtually on a continuation of the broad northeasterly trending belt of mineralised basement (Pan African) pegmatites which contain economic amounts of cassiterite and columbite–tantalite, and which are most abundant just to the south-west of the Jos Plateau (Sec. 6.5). The Nigerian Younger Granites are thus most abundant at the intersection of this 'pegmatite belt' and the projection of the north–south line of separation between the southern halves of Africa and South America.

Fractures and faults in the basement must have influenced the location of individual Younger Granite complexes, but such fractures are neither numerous nor obvious on the Jos Plateau, and it is not easy to perceive how they controlled emplacement of the granites. In the Aïr, on the other hand, where the Younger Granites were intruded into a crustal 'swell' between the Iullmedden and Chad Basins, numerous faults and shear zones have been mapped in the Pan African basement, and these must have helped to control emplacement of the complexes.

Gravity surveys over the Nigerian Younger Granites have shown that the Jos Plateau area is characterised by a broad regional negative gravity anomaly, elongated in a NW direction, similar in trend to the Bida Basin and Niger valley. Individual complexes are associated with large local negative anomalies, interpretation of which suggests that the complexes extend to depths of 10–12 km and retain their identities to these depths, that is, they are unlikely to be connected to a larger 'parent' granite body deep in the crust.

15.4 Petrogenesis

The regions of crustal doming into which the Younger Granites were emplaced are similar to those associated with rift valley formation and related igneous activity, of which the best examples are in eastern Africa. Such regions are underlain by mantle that is warmer and less dense than average. Pressure relief associated with the upwards doming may help to initiate the partial melting of this anomalous mantle, typically generating magmas of the olivine basalt–trachyte–phonolite association with geochemically alkaline to peralkaline affinities. This kind of magmatism characterises the non-orogenic igneous activity of continental swells and rifts all over the world.

The Younger Granite province differs from most other alkaline–peralkaline associations in some important respects. Most important, rhyolitic and granitic compositions greatly predominate over basic and intermediate varieties, whereas they are normally subordinate. The basic and syenitic rocks do not show strong alkaline tendencies and undersaturated varieties (phonolites and nepheline syenites) are not recorded; moreover, only some of the rhyolites and granites are peralkaline. Nor is there any major contemporaneous tensional faulting or rifting. The granites were not fractionation products

of large basic magma chambers emplaced at mid-crustal depths beneath them. The crustal uplifts and the regional and local negative gravity anomalies associated with the complexes preclude the existence of the huge volumes of basic cumulates that would be left after production of the rhyolites and granites by magmatic differentiation.

The magmas must have been generated by partial melting in the upper mantle and lower crust beneath the crustal domes into which the complexes were emplaced. Isotopic data imply significant crustal involvement in the genesis of the magmas. Crustal melting may have been facilitated by volatile fluxing from the upper mantle. As the magmas rose towards the surface, they were modified to varying degrees by further crustal melting and assimilation in the warm and ductile lower crust. As they rose into the brittle upper crust, further ascent was by fracturing and stoping (cauldron subsidence). Crystal fractionation may also have occurred, to contribute to the diverging peralkaline and peraluminous geochemical trends. The distribution patterns of rare-earth elements in some peralkaline granites, chiefly a strong europium (Eu) depletion, are consistent with precipitation of early plagioclase feldspar and with crystallisation of late alkali-bearing amphiboles from volatile-rich low-temperature fluids.

The gabbroic and syenitic rocks probably represent relatively unmodified upper mantle melts, and some of the peralkaline granites could also have a significant mantle component. The fayalite- and pyroxene-bearing porphyries and related rhyolites solidified from comparatively anhydrous magmas that could have come mainly from the lower crust. The more aluminous biotite granites are likely to have formed from magmas that experienced greater modification by assimilation at higher crustal levels than the others, and this would be consistent with their generally later emplacement in the intrusion sequence. The earlier magmatism must have contributed to an overall warming of the crust, so that crustal melting became progressively easier with time. The biotite granites are also generally more hydrous and are associated with much of the late-stage deuteric and hydrothermal activity, notably albitisation and fluorine enrichment, greisen formation and mineralisation.

There has been some debate about the source of the mineralisation, associated with the late-stage albitising fluids, centring on the question of a crustal versus a mantle source for the tin and niobium. In Nigeria, the close spatial association of Younger

Granites and Older Granite pegmatites, both characterised by mineralisation dominated by tin and niobium–tantalum minerals, has been used to support both arguments. The metals could have come from a common deep source at different times or they could have been remobilised from the crust during the Younger Granite episode. However, the fact that the Younger Granites retain their identity down to depths of 10–12 km and that they were manifestly emplaced into cold and brittle crust argues against any significant remobilisation of older rocks at depths of less than about 10 km. Of possible additional relevance to the debate is the fact that tourmaline is abundant in the basement and Older Granites, but is not recorded from any of the Younger Granites.

15.4.1 Younger Granites and mantle plumes

Plate movements are intermittent, periods of motion alternating with stationary episodes which may last for tens of millions of years. It has been proposed that, during these stationary episodes, thermal plumes or hot spots in the upper mantle can focus sufficient heat on the overlying lithosphere and crust to cause doming and igneous activity. The plumes are supposed to have a fixed position in the mantle, so that they should leave a record of plate movements over them.

The somewhat irregular decrease in age of Younger Granites from north to south in Nigeria has led to the proposition that the African plate was moving northwards over a mantle plume during the Jurassic, but sufficiently slowly for magmas of the different complexes to be generated in turn. Earlier manifestations of this same plume have been correlated with Younger Granites of the Aïr and southern Niger. These are identified as resulting from plume activity during stationary episodes 340–290 Ma and 430–380 Ma ago respectively. After the Jurassic the African plate moved further north, and it is further proposed that about 100 Ma ago the plume lay beneath the triple junction now occupied by the Niger Delta and thus was a major factor in opening the South Atlantic and forming the Benue Trough.

The idea is superficially attractive, but neither the rate nor direction of plate movements deduced from the Younger Granites is consistent with the Palaeozoic–Mesozoic motions determined by palaeomagnetism. Moreover, in northeastern Africa, in Egypt and the Sudan, there are over 100 anorogenic ring complexes of Younger Granite type, ranging in age from over 500 Ma to less than 100 Ma

and distributed over a vast area. These display no discernible age trends and fixed mantle plumes cannot have been responsible for their emplacement. Otherwise they would provide an excellent record of movements of the African plate over a period of some 450 Ma.

15.5 Economic potential of the Younger Granites

The Nigerian Younger Granites have long been famous for their *tin* mineralisation. The primary mineralisation is almost entirely associated with the biotite granites and in particular with those varieties that have been subjected to late-stage albitisation. Greisen and quartz veins and silicified fractures carry the cassiterite, and these form stockworks in the roof zones of albite–biotite granites, occasionally extending a short distance into mildly metasomatised country rocks. The richest alluvial concentrations occur in the vicinity of tin-bearing granites that have undergone only shallow erosion. Most of Nigeria's tin production has come from alluvial concentrations derived from an area within a radius of about 50 km of Jos itself. Cassiterite is only rarely mined from primary sources.

Wolframite also occurs in the primary veins along with the cassiterite, but only in small quantities, and production of this mineral all but ceased in the mid–1960s. Other minerals include topaz, fluorite and sulphides, especially sphalerite and chalcopyrite, as well as pyrite and occasional molybdenite (see also below). It has been known for many years that *beryl* is another constituent of the veins, but not until the 1980s did it become apparent that the Younger Granites are hosts to gem quality crystals of this mineral, in yellow and white varieties as well as *aquamarines* and even *emeralds*.

The alluvial concentrations of cassiterite are placer deposits among gravel pockets in both ancient and modern stream channels. Grades may be as high as 3 lb per cubic yard (about 1.5 kg m^{-3}), but are usually nearer to one-tenth of this. The alluvial overburden is not normally much more than about 30 or 40 m, and this normally presents relatively few problems of stripping and mining. However, where late Tertiary to Quaternary basalt lavas (Ch. 16) flowed down stream channels to form a hard capping to the underlying alluvial deposits, both exploration and extraction are a great deal more difficult.

Production of cassiterite was of the order of 10 000 tonnes annually for many years, and was the mainstay of Nigeria's mineral industry, until overtaken by oil. Production declined during the 1970s, to a little over 3800 tonnes in 1979, and was less than 3200 tonnes in 1981, falling to about 1700 tonnes in 1983. Virtually all the cassiterite was exported as concentrates until 1962, when a tin smelter was commissioned at Jos. A small proportion of the tin metal produced is used locally, the rest being exported.

The second most valuable mineral associated with the Younger Granites is *columbite*, also found in alluvial concentrations. It occurs as an accessory mineral in the albitised biotite granites, and so its distribution in placer deposits is not very different from that of the cassiterite. However, it seems to occur mainly in the modern drainage channels, being virtually absent from older alluvial deposits. Where they occur together the ratio of columbite to cassiterite in placer deposits is of the order of 1 : 20, but it can be as high as 1 : 3. Production of columbite is much less than that of cassiterite, and fluctuated between about 1000 and 2000 tonnes annually until the mid-1970s, after which it declined considerably, to less than 400 tonnes in 1981, falling still further in 1982 and 1983.

Total reserves of cassiterite and columbite in the Nigerian Younger Granite province are estimated to be of the order of 140 000 and 70 000 tonnes respectively, sufficient for several years of further production at existing rates. When the alluvial reserves are exhausted, the greatest potential for mining primary tin and niobium ores may be in those Younger Granite complexes where the mineralised stockworks are still unexposed beneath volcanic rocks of the calderas. Other minerals found among the alluvial concentrates include *magnetite*, *ilmenite*, *zircon*, *thorite* and *monazite*, none of which is presently of economic interest, but could prove to be useful in future. Gem quality *beryl* is also present, and there may also be a market in semi-precious gem quality *topaz*, which is also plentiful among the alluvial concentrates. Some of the rarer accessories among the other varieties of Younger Granite, such as *pyrochlore* and *cryolite*, though presently of no value, could one day also become worth extracting.

Molybdenum has been mined on a small scale from one of the Younger Granites west of Jos. It occurs as disseminated molybdenite in a peralkaline (arfvedsonite–aegirine) granite, but production has never exceeded a few tonnes annually.

Cassiterite is extracted from alluvial concentrations derived from both Aïr and Cameroun Younger Granites, but production is very small, of the order of 100 and 35 tonnes per year respectively.

There are plans to increase exploration and output, but the prospects are not promising. The area occupied by biotite granites (the mineralised host rocks) is less than 5% of the total Younger Granite outcrop in Niger, and less than 15% in Cameroun, as against about 55% in Nigeria.

On the other hand, the Younger Granites have provided the likely source of the rich uranium deposits in Niger (Sec. 12.1.3), and similar deposits could occur in Cameroun also. The most likely hosts for secondary uranium mineralisation in Nigeria would seem to be the Cretaceous–Tertiary sequences of the Benue Trough, but so far no significant deposits have been discovered there.

Bibliography

Ajakaiye, D. E. 1976. A gravity survey over the Nigerian Younger Granite Province. In *Geology of Nigeria*, C. A. Kogbe (ed.). Lagos: Elizabethan Press.

Ajakaiye, D. E. and K. Burke 1973. A Bouguer gravity map of Nigeria. *Tectonophysics* **16**, 103–15.

Black, R. and M. Girod 1970. Late Palaeozoic to Recent igneous activity in West Africa and its relationship to basement structure. In *African magmatism and tectonics*, T. N. Clifford and I. G. Gass (eds). Edinburgh: Oliver & Boyd.

Bonin, B., P. Bowden and Y. Vialette 1979. Le comportement des éléments Rb et Sr au cours des phases de minéralisation: l'exemple de Ririwai (Liruei), Nigeria. *C.R. Acad. Sci. Paris*, D **289**, 707–10.

Bowden, P. and J. A. Kinnaird 1978. Younger Granites of Nigeria – zinc-rich tin province. *Trans. Inst. Mining Metall. B* **87**, 66–9.

Bowden, P. and J. E. Whitley 1974. Rare-earth patterns in peralkaline and associated granites. *Lithos* **7**, 15–21.

Bowden, P., J. E. Whitley and O. van Breemen 1976. Geochemical studies on the Younger Granites of northern Nigeria. In *Geology of Nigeria*, C. A. Kogbe (ed.). Lagos: Elizabethan Press.

Bowden, P., O. van Breemen, J. Hutchinson and D. C. Turner 1976. Palaeozoic and Mesozoic age trends for some ring complexes in Niger and Nigeria. *Nature* **259**, 297–9.

Bowden, P., J. N. Bennett, J. A. Kinnaird, J. E. Whitley, S. I. Abaa and P. K. Hadzigeorgiou-Stavrakis 1981. Uranium in the Niger–Nigeria Younger Granite province. *Mineral Mag.* **44**, 379–88.

Cahen, L., and N. J. Snelling 1984. Geochronology and evolution of Africa. Oxford: Oxford University Press.

Falconer, J. D. 1911. *The geology and geography of northern Nigeria*. London: Macmillan.

Gazel, J., M. Lasserre, J-C. Limasset and M. Vachette 1963. Ages absolus des massifs granitiques ultimes et de la minéralisation en étain du Cameroun central. *C.R. Acad. Sci. Paris* **256**, 2876–8.

Harris, N. B. W. and G. F. Marriner 1980. Geochemistry and petrogenesis of a peralkaline granite complex from the Midian Mountains, Saudi Arabia. *Lithos* **13**, 325–37.

Ike, E. C. 1983. The structural evolution of the Tibchi ring-complex: a case study for the Nigerian Younger Granite Province. *J. Geol. Soc. London* **140**, 781–8.

Imeokparia, F. G. 1982. Geochemical relationships to mineralization of granitic rocks from the Afu granite complex, central Nigeria. *Geol. Mag.* **119**, 39–56.

Jacobson, R. R. E., W. N. MacLeod and R. Black 1958. *Ring-complexes in the Younger Granite province of northern Nigeria.* Geol. Soc. Lond. Memoir No. 1.

Jacquemin, H., S. M. F. Sheppard and P. Vidal 1982. Isotopic geochemistry (O, Sr, Pb) of the Golda Zuelva and Mboutou anorogenic complexes, north Cameroun: mantle origin with evidence for crustal contamination. *Earth Planet. Sci. Lett.* **61**, 97–111.

Karche, J-P. and C. Moreau 1977. Note préliminaire sur le massif subvolcanique à structure annulaire d'Abontorok (Aïr–Niger). *C.R. Acad. Sci. Paris D* **284**, 1259–62.

Karche, J-P. and M. Vachette 1976. Migration des complexes subvolcaniques à structure annulaire du Niger. Consequences. *C.R. Acad. Sci. Paris D* **282**, 2033–6.

Kinnaird, J. A. (ed.) 1981. *Geology of the Nigerian anorogenic ring complexes* 1:500000. St Andrews University Younger Granite Research Group. Edinburgh: Bartholomew & Sons.

Kogbe, C. A. and A. U. Obialo 1976. Statistics of mineral production in Nigeria (1946–1974) and the contribution of the mineral industry to the Nigerian economy. In *Geology of Nigeria*, C. A. Kogbe (ed.). Lagos: Elizabethan Press.

Olade, M. A. 1980. Geochemical characteristics of tin-bearing and tin-barren granites, northern Nigeria. *Econ. Geol.* **75**, 71–82.

Schuiling, R. D. 1967. Tin belts on the continents round the Atlantic Ocean. *Econ. Geol.* **62**, 540–50.

Sillitoe, R. H. 1974. Tin mineralisation above mantle hot spots. *Nature* **248**, 497–9.

Turner, D. C. 1976. Structure and petrology of the Younger Granite ring complexes. In *Geology of Nigeria*, C. A. Kogbe (ed.). Lagos: Elizabethan Press.

Turner, D. C. and P. Bowden 1979. The Ningi-Burra complex, Nigeria: dissected calderas and migrating magmatic centres. *J. Geol. Soc. Lond.* **136**, 105–20.

Turner, D. C. and P. K. Webb 1974. The Daura igneous complex, N. Nigeria; a link between the Younger Granite districts of Nigeria and S. Niger. *J. Geol. Soc. Lond.* **130**, 71–8.

UNESCO 1968. *Proceedings of the symposium on the granites of West Africa (1965).* UNESCO Natural Resources Research Series, No. VIII. Paris: UNESCO.

Vail, J. R. 1976. Location and geochronology of igneous ring-complexes and related rocks in north-east Africa. *Geol. Jahrb. B* **20**, 97–114.

van Breemen, O., J. Hutchinson and P. Bowden 1975. Age and origin of the Nigerian Mesozoic Granites: a Rb–Sr isotopic study. *Contrib. Mineral. Petrol.* **50**, 157–72.

Wright, J. B. 1970. Controls of mineralization in the older and younger tin fields of Nigeria. *Econ. Geol.* **65**, 945–51.

Wright, J. B. 1973. Continental drift, magmatic provinces and mantle plumes. *Nature* **244**, 565–7.

16 *Cretaceous and Cenozoic magmatism*

SUMMARY

Kimberlite dykes and pipes found in many parts of the West African craton are probably mainly of Cretaceous age, but some may be much older. Several are diamondiferous. A suspected kimberlite in Nigeria is probably a basaltic explosion vent.

In the Benue Trough there is a pre-Cretaceous suite of mainly acid igneous rocks. During the Upper Cretaceous mainly basic rocks were erupted and intruded among the sediments of the trough.

Cenozoic volcanism is of alkaline affinities and is virtually confined to Pan African areas east of the craton. On the Jos Plateau early Cenozoic(?) volcanism produced the Fluvio-Volcanic Series of intercalated sediments and basaltic lavas. Scattered Plio-Pleistocene volcanism dominated by basaltic cones and plateaux, with minor phonolite and trachyte plugs, characterised the Hoggar and Aïr regions and also the Jos Plateau and Benue Trough. Many basalts contain megacrysts and upper mantle inclusions. The Cameroun line forms the major part of the Gulf of Guinea province, with large shield volcanoes and basalt plateaux, also plugs of phonolite, trachyte and rhyolite. There was minor Cenozoic volcanism in the Dakar region of Senegal.

Most of the Cenozoic magmatism occurs in long-established areas of crustal doming or uplift and cannot realistically be related to fixed hot spots in the mantle.

There are several rich alluvial diamond fields on the craton, some of which can be traced to kimberlites. The economic potential of the Cenozoic volcanic rocks is confined to their use as roadstone or other aggregate, although alluvial tin is found among some of the Fluvio-Volcanics on the Jos Plateau, and the clays are used for ceramics. Some of the basalt plateaux have aquifers of limited size. Small amounts of sapphire occur in alluvial deposits at one locality.

16.1 Igneous activity in the Cretaceous

The kimberlites and related rocks emplaced into the West African craton may well have had an extended history that only terminated in the Cretaceous, about 100 Ma ago, but it is convenient to treat them here. The early acid igneous rocks of the Benue Trough are provisionally assigned to the Cretaceous, though radiometric dating may show them to be older. Only the magmatic activity that actually accompanied development of the Benue Trough can be placed with reasonable confidence in the Upper Cretaceous.

16.1.1 Kimberlites and related rocks

Kimberlites occur in many places on the West African craton, as dykes, small pipes and breccia zones. On a local scale, kimberlite dykes may fill existing fractures between sets of regionally developed meridional (N–S) faults and shear zones of probable Archaean or Proterozoic age. The dyke sets have variable trends that do not in general coincide with those of the basement or the Mesozoic dolerites. The dykes may be in *en echelon* arrangement, and pipes develop where dykes are cut by fractures. In Ivory Coast, the kimberlite dykes form part of a swarm containing dykes of other alkaline rock types.

The kimberlite is usually a porphyritic to fine-grained grey to gre n rock, consisting mainly of serpentine, phlogopite, olivine, pyroxene and carbonate, with magnesian garnet and ilmenite megacrysts, and often well rounded xenoliths of crustal and mantle rocks. Late-stage veins of calcite and serpentine are common, and there has been some **fenitisation** of the basement wall rocks. Many of the kimberlites are diamondiferous. Kimberlites are easily weathered rocks and are therefore not easily found and identified. There are some important alluvial diamond fields in West Africa (Sec. 16.4) for which the source kimberlites have not yet been located. Moreover, not all authorities agree about the identification of those kimberlites that have been reported.

Kimberlites are usually emplaced as **diatremes**. They are gas-rich magmas that rise rapidly from great depths, 'drilling' their way explosively into the upper crust by a combination of gas fracturing, stoping and fluidisation. Kimberlite is the only igneous rock in which diamond is found. Diamond forms at pressures appropriate to depths of 150 km or more, and only survives because it is transported to the surface very rapidly.

K/Ar dating of kimberlites in Sierra Leone has indicated ages of between about 80 and 140 Ma, consistent with geological evidence for emplacement during the Cretaceous. However, kimberlites both further north and further east give much older dates: around 1070 Ma in Mali and 1430 Ma in Ivory Coast. These are Proterozoic ages, though clearly much younger than the Eburnian (Sec. 4.5). Nonetheless, the dates may not be valid because daughter isotopes are thought to accumulate in kimberlite magmas prior to eruption, thus making the rock appear radiometrically much older than its age of emplacement. There is evidence that these unique magmas reside for a considerable time at great depths within the lithosphere, prior to their rapid and explosive ascent to the surface. Emplacement of kimberlites should therefore be facilitated by the formation of deep lithospheric fractures such as those formed during plate adjustments of the kind that must have accompanied the break-up of Gondwanaland in the Mesozoic. On the other hand, the later Proterozoic ages for Mali and Ivory Coast kimberlites could well be valid. If kimberlites could be emplaced in Archaean crust in the early Proterozoic, to supply diamonds for the Birimian of Ghana and Ivory Coast (Sec. 4.7.3), there is no *a priori* reason why they should not have been emplaced at intervals since then. To argue that they are all Cretaceous is also to imply a residence time in the lithosphere of 10^9 years or so, which seems excessive. Moreover, there is evidence that some of the diamonds in northern Liberia and adjacent Guinea come from graphitic schists of Archaean age.

Attempts have been made to relate the kimberlite fields of West Africa to continental extensions (or progenitors) of oceanic fracture zones on the one hand, and to movements of lithosphere over mantle plumes on the other. All that can be stated with confidence, however, is that in West Africa, as elsewhere, kimberlites occur along major crustal fractures and they occur mainly in crust that is Lower Proterozoic or older.

A suspected kimberlite in northwestern Nigeria satisfies the first of these two criteria, but fails on the second, for the basement rocks are Pan African. The circumstantial evidence is quite strong: magnesian garnet and ilmenite in stream sediments were used to locate the suspected source of at least two alluvial diamonds found in the region, and geophysical

investigations have demonstrated the existence of a pipe-like body extending to at least 0.5 km depth and with a diameter of about 150 m. Nonetheless, unequivocal petrographic evidence of actual kimberlite is still lacking.

No phlogopite has been found, and the magnesium content of the ilmenite is much lower than that typical of kimberlites (4% as against 6% or more). The rocks are much weathered, there is thick laterite cover, and such fragments as can be identified are of basaltic aspect. The occurrence is more likely related to a very minor, scattered and little known outbreak of basaltic volcanism of probably Jurassic age in northwestern Nigeria. The source of the Nigerian diamonds is still not unequivocally identified.

Other alkaline rocks Three small isolated centres of alkaline igneous rocks occur in the western part of the region. They are the Isles de Los complex of coastal Guinea, and the Songo and Bagbe complexes of Sierra Leone, situated respectively just east of the Freetown intrusion (Sec. 14.1.2) and due south of the two major diamond fields (Sec. 16.4). All three are probably ring complexes and they are dominated by nepheline syenites, associated with pyroxenites, carbonatites and phosphate-rich rocks in various proportions. The Bagbe complex is surrounded by fenitised granitic basement. All the complexes are thought to be coeval and to be part of the kimberlite province. The Songo complex has been dated at 150 ± 10 Ma.

16.1.2 Cretaceous magmatism of the Benue Trough

There was minor but significant igneous activity in several parts of the Benue Trough during the Cretaceous, related to the tectonic evolution of the trough. The oldest igneous rocks comprise an assemblage of granite and porphyry intrusions, andesitic, rhyolitic and basaltic lavas and agglomerates, and felsite dykes. The rocks cut and overlie basement of the Zambuk Ridge south-west of the Biu Plateau (Fig. 11.1) and they are collectively called the **Burashika Group**. The age of the Burashika Group is unknown, but rhyolites cutting basement near Gboko (north-east of Abakaliki) have been dated at 113 Ma and may be part of the same episode.

In the upper Benue Trough, where exposures are generally fairly good, thin lavas and tuffs have been found intercalated with most of the Cretaceous sedimentary units, showing that sporadic minor volcanism occurred throughout the whole period of Cretaceous sedimentation in this region at least. Individual volcanic layers are rarely more than a metre or so thick and never exceed 5 m. The rocks are almost invariably altered to red or green clays, presumably because they were erupted among layers of wet sediments. The rocks were originally basaltic in composition, and their tectonic setting in a rifted environment suggests that they had alkaline affinities typical of such rocks in similar settings elsewhere – but they are too altered for this to be verified.

Extrusive rocks appear to be less common in the middle and lower Benue Trough, but thin lavas and tuffs would be more difficult to detect in these rather poorly exposed regions, especially if altered to clays. In the Abakaliki area, pyroclastic breccias and some altered basalts with geochemically alkaline affinities have been interpreted as belonging to an early Albian volcanic episode.

No intrusive rocks of Cretaceous age have so far been found in northeastern parts of the Benue Trough, but minor intrusions increase in number and diversity towards the south-west, being most abundant in the region of the Abakaliki anti-clinorium (Fig. 11.1). The intrusions are mostly dykes and sills, but small stocks and bosses a few kilometres in diameter also occur. The commonest rocks are dolerites, but syenites and diorites are also recorded, and undersaturated rock types have been identified, including teschenite, tinguaite and nepheline syenite. So at least part of the Cretaceous igneous province of the Benue Trough has alkaline affinities. However, as the dolerites frequently contain orthopyroxene and pigeonitic clinopyroxene, at least some of the magmatism was of tholeiitic type. The intrusive rocks are commonly altered (see below), and rocks identified as diorites may originally have been gabbros in which the feldspars were albitised by hydrothermal alteration.

From the rather limited evidence available, it seems that whereas volcanic activity continued throughout the Cretaceous infilling of the Benue Trough, the intrusions are most abundant in Albian sediments, less common in Turonian sediments and absent from post-Turonian sediments. As the greatest concentration of intrusions lies in the Abakaliki anticlinorium, it is likely that their emplacement was structurally controlled. The climax of intrusive activity probably occurred during the Santonian folding episode in the lower Benue Trough (Table 11.1).

The Cretaceous igneous activity and the lead–zinc mineralisation in the Benue Trough (Sec. 12.2.2) are indirectly related in that both are the products of a linear zone of increased heat flow beneath the Benue Trough during its evolution. The same heat that gave rise to the magmatism also powered the hydrothermal circulation which deposited the lead–zinc ores and was responsible for much of the alteration in the intrusive rocks.

16.2 Cenozoic volcanism

Volcanic activity dominated by magmas of the continental alkali olivine basalt–trachyte–phonolite association commenced in many parts of Africa about 35 Ma ago. It increased in intensity during the Miocene and Pliocene and then declined, though it has persisted in a few places to the present day (e.g. East Africa, Mount Cameroun). It is virtually confined to non-cratonic (Pan African) areas of basement uplift ('swells') that border the sedimentary basins and are sometimes associated with rifting.

Figure 13.1b summarises the distribution of Cenozoic volcanics in West Africa. In general, basaltic lavas predominated, forming important plateau accumulations and broad shield volcanoes in some places, in others only small cinder cones and thin flows. The more viscous trachytes and phonolites commonly formed plugs and domes or thick flows (tholoids). These make impressive topographic features because of their resistance to erosion and they may dominate the landscape despite their much smaller volume relative to the basalts.

16.2.1 Hoggar and southern Aïr

In both these regions considerable epeirogenic uplift of the basement occurred before and during volcanism, of the order of 1000 m in the Hoggar and about 500 m in southern Aïr. Both regions are cut by faults and shears, and in southern Aïr ring fractures associated with the Younger Granites provided additional controls on the location of eruptive centres.

A few small ring complexes with diorites, syenites and nepheline syenites lie north-east of Atakor (Fig. 13.1b) and may be Cretaceous or older. Cenozoic volcanism in the Hoggar began with Miocene plateau basalts, which erupted intermittently, for there are interflow weathering horizons marked by alteration to clays and lateritic soils. Numerous Miocene plugs and domes (tholoids) of phonolite and trachyte give rise to spectacular landscapes in this region. Quaternary volcanism was confined mainly to eruption of basalts, in the form of cinder cones built of ash, scoria and volcanic bombs, and relatively thin flows that filled pre-existing valleys. Generally similar activity occurred in southern Aïr, though much less abundant and without an initial plateau basalt stage. Many of the basaltic rocks contain ultramafic (peridotite) nodules derived from the underlying upper mantle.

16.2.2 The Gulf of Guinea province

Basalt lava plateaux, trachyte plugs and domes, large central volcanoes and small basalt cinder cones with thin flows are all found among the more southerly manifestations of Cenozoic volcanism in West Africa (Figs. 11.1 and 13.1b).

This province also includes the remarkable offshore continuation of the Cameroun volcanic line, the four islands situated in the Gulf of Guinea itself. Areas of basement doming include the Jos Plateau, which rose nearly 1000 m during the Cenozoic; the Bamenda Highlands and Mambilla Plateau, southeast of the Benue Trough, with probably still greater uplifts; and the Adamaoua Highlands further east, where lavas of the Ngaoundere Plateau overlie the Ngaoundere Fault zone, which was reactivated in the Cretaceous (Secs 8.4 and 11.1.1). The flood basalts of the Biu Plateau are situated on the Zambuk Ridge of the upper Benue Trough, and the smaller Longuda Plateau lies near the bifurcation of the upper Benue Trough, which has been identified as a possible secondary triple junction (Sec. 11.1).

On the Jos Plateau, the earliest eruptions were basaltic lavas of the somewhat enigmatic **Fluvio-Volcanic Series**. These deposits form laterite-capped sheets and residual flat-topped hills at elevations varying from about 1100 to 1400 m, and consist of fluvio-lacustrine gravels, sands and clays, interbedded with or overlain by yellow and purple clays representing kaolinised and bauxitised basalt lavas. The laterite capping to these beds has been correlated with the laterite developed on the Palaeocene Kerri Kerri Formation of the upper Benue Trough (Table 11.1), so the Fluvio-Volcanic Series may also be of Lower Tertiary age. The sediments and volcanics reach a maximum thickness of 300 m and were deposited in depressions on an ancient landscape of moderate relief, burying all but the highest hills. These are of some economic interest, for the basal gravels are frequently rich in alluvial cassiterite (Sec. 16.5).

Basalts post-dating the Fluvio-Volcanics on the Jos Plateau have been divided into Older and Newer Basalts on the somewhat subjective criteria of relative degrees of weathering and lateritisation. Radiometric dating suggests that they are no older than 3 Ma, and there was probably more or less continuous activity from late Pliocene to very recent times. Some of the cinder cones and their thin valley-filling flows look very young, and there are all gradations to heavily eroded cones and to lava sheets and remnants without a discernible source. Several cones are aligned along presumed basement fracture trends. The youngest cones contain abundant ultramafic nodules and megacryst phases of upper mantle origin. Rounded fist-sized fragments of diopsidic pyroxene and magnesium (pyropic) garnet are especially striking. Other minerals include magnesium ilmenite, alkali (kaersutitic) amphibole and sodic plagioclase. The most unusual megacryst phases occur in extensive basalt flows immediately south of the Jos Plateau, which contain large numbers of zircon and corundum crystals, many of gem quality (Sec. 16.5).

A single plug of trachytic phonolite on the Jos Plateau probably pre-dates the basalts. It is of particular interest for its rich content of upper mantle and crustal inclusions (peridotite, gabbro, anorthosite, syenite, granite) and large megacrysts of anorthoclase feldspar.

In the middle and upper Benue Trough and the Yola arm, plugs and domes (tholoids) of trachyte and phonolite form characteristically steep-sided hills. Samples of these rocks have given ages of between 12 and 22 Ma, placing them in the Miocene.

In the Benue Trough basaltic volcanism probably did not begin until the end of the Miocene, when the Biu and Longuda Plateaux were built. The bulk of the Biu Plateau consists of Pliocene basalts, erupted from small vents or fissures and spreading thinly over wide areas. There are numerous small pyroclastic cones within the sequence, interpreted as the result of explosive eruptions, caused by ground water percolating into the growing lava pile and coming in contact with fresh lava. There was a break in activity at the end of the Pliocene, with weathering, erosion and laterite formation. Activity resumed in the Quaternary, in the form of thin valley-filling basalt flows from small cinder cones, some of which have a very youthful aspect. These cones are also notable for their abundant inclusions of upper mantle peridotite and plentiful megacrysts of minerals similar to those found in Quaternary basalts on the Jos Plateau.

The chronology of the Biu Plateau is well controlled by radiometric dating, samples of the plateau basalts falling in the 7–2 Ma range, the Quaternary basalts giving 1 Ma or less. The smaller Longuda Plateau to the south-west has not been dated, but its geomorphology suggests that it is also built of Pliocene basalts, though there are no later Quaternary cones here. In contrast, the small Song field to the south-east of the Biu Plateau consists entirely of Quaternary flows and cones aligned WNW–ESE, parallel with the boundary between basement and Cretaceous sediments on the northern side of the Yola arm of the Benue Trough.

There are great numbers of basaltic necks and plugs and flow remnants among the Cretaceous sediments of the middle and upper Benue Trough, extending into the Yola arm. Although their physiographic aspect might suggest that they are similar in age to the phonolite and trachyte bodies, radiometric dating of some of them has given Pliocene ages, so they are contemporaneous with growth of the Biu and Longuda Plateaux.

The Cenozoic volcanics of the Benue Trough can almost invariably be distinguished from the Cretaceous igneous rocks by their fresh and unaltered aspect, both in outcrop and in thin section.

The Cameroun line The main **Cameroun line** extends from the islands in the Gulf of Guinea through Mount Cameroun and the Bamenda Highlands, and across the Yola arm of the Benue Trough towards Lake Chad. It is defined both by Lower Tertiary Younger Granite complexes and by the Cenozoic volcanics. An eastern branch diverges north of the Bamenda Highlands and extends to the Ngaoundere Plateau of the Adamaoua Highlands, which lies on the Ngaoundere Fault Zone (Fig. 11.1).

Volcanism was sparse along the northern third of the line, where domes, plugs and necks of phonolite, trachyte and basalt may belong with the Miocene and Pliocene episodes of the Benue Trough.

The oldest lavas on the Bamenda Highlands are basalts that have been altered to multicoloured clays resembling those of the Fluvio-Volcanic Series of the Jos Plateau. They cannot be much older than Miocene, however, for Younger Granites of the Cameroun line have dates as young as 35 Ma and are separated from the basaltic volcanics by a period of erosion. Most of the volcanic activity in Cameroun was probably Pliocene, producing mainly basalts, but also flows and plugs of trachyte, rhyolite and subordinate phonolite in the Bamenda Highlands and the Ngaoundere Plateau. There are large basaltic and trachyte–rhyolite shield volcanoes with summit calderas in the Bamenda Highlands and basalt–trachyte volcanism characterises the Manenguba Mountains to the south.

Signs of recent basaltic activity in several places show that volcanism persisted through the Quaternary, and Mount Cameroun is still active. It is the largest mountain in West Africa, rising some 4000 m above sea level, built mainly of basaltic lavas and ashes, with cinder cones scattered about its flanks. The last significant eruption was in 1922, when lavas emanated from small cones on the upper slopes and reached the sea on the west side of the mountain. The most recent eruptions were in 1954, 1959 and 1982.

The adjacent offshore island of Bioko (formerly Fernando Po) is also built mainly of basalts, probably mostly of Pliocene age. The more distant offshore islands are also formed mainly of basalt flows, overlain and intruded by phonolites and trachytes, sometimes followed by more basalts. Radiometric dates and field relationships suggest that volcanism commenced as long ago as the late Oligocene (*c.* 30 Ma), but at least some of the phonolites and trachytes are Pliocene (3–5 Ma).

There is no obvious systematic pattern of petrographic variation in time or space either along the Cameroun line or within the Gulf of Guinea province as a whole. This must be taken into account when considering the geotectonic setting.

There is a striking resemblance between the distribution of the main Pliocene-Quaternary volcanics in the Gulf of Guinea province and the Y-shape of the Benue Trough, and the two patterns can be superimposed (Fig. 16.1). When the South Atlantic opened around 100 Ma ago, the Benue Trough originated as the failed arm of the Gulf of Guinea triple junction. A zone of crustal attenuation and elevated heat flow developed over a linear plume in the underlying asthenosphere, bifurcating in the north-east to form the Yola branch of the trough. Over this zone, the crust was stretched and rifted, and the linear sedimentary basin developed, accompanied by some igneous activity.

Analysis of magnetic anomaly patterns, transform faults and bathymetry in the South Atlantic has been used to suggest that, about 80 Ma ago, the African plate was rotated clockwise slightly. It is proposed that as a result the linear plume in the asthenosphere came to lie south-east of the Benue Trough and from

Figure 16.1 The main volcanic areas of the Gulf of Guinea province superimposed on the outline of the Benue Trough.

its new position it subsequently generated the Cenozoic volcanism of the Cameroun line.

The suggestion has considerable appeal when only the major Cenozoic volcanism is considered. However, the Cameroun line itself almost bisects the Y-shape of Figure 16.1, extending well into northwest Cameroun, where it is represented by both Tertiary Younger Granites and Cenozoic volcanics. Moreover, there was late Cretaceous igneous activity in the Benue Trough and Cenozoic volcanism there was not confined to the Biu and Longuda Plateaux. There was Cenozoic volcanism on the Jos Plateau as well. On the larger scale, other difficulties become evident. The final phase of movement of the African plate was associated with opening of the South Atlantic and did not end until about 40 Ma ago. By then, what is now Cameroun would have been hundreds of kilometres away from any linear plume that had been established beneath the Benue Trough, yet the major volcanism of the Cameroun line did not commence until the Miocene, around 25 Ma ago. The correspondence illustrated in Figure 16.1 may appear too close to be fortuitous, but it cannot be explained by movements of the African plate over fixed hot zones in the mantle.

The crustal uplifts in Cameroun must be at least as long-lived and persistent as that of the Jos Plateau, for example. The main volcanic occurrences in Cameroun lie close to the great Ngaoundere Fault

Zone, which was reactivated in the Cretaceous and provides a further potential control on the location of Cenozoic igneous activity in this region.

16.2.3 Cenozoic volcanism west of the craton

The small outbreaks of Cenozoic volcanic activity west of the craton appear to be confined to the vicinity of the Cape Verde peninsula (Fig. 13.1b). There are flows of alkali basaltic lavas and tuffs containing blocks of Miocene limestones, and forming the southern part of the peninsula, and they may be of late Tertiary age. Basaltic flows and pyroclastics of the dissected Manelles volcano form the northeastern part of the peninsula. They overlie Pleistocene sands and laterites and are therefore of very recent age. The Dakar volcanics are practically in an oceanic setting, and it seems likely that they are related to the volcanic islands of the Cape Verde group, which lie a few hundred kilometres west of Dakar. Other volcanics occur locally in the succession, highly altered to clay minerals. Some 80 km east of Dakar there are basanitic plugs with concentric structures and coarse-grained (gabbroic) cores. The bodies are a few hundred metres across and they were probably emplaced along deep-seated meridional faults some time during the Oligo-Miocene.

16.3 Controls of Cenozoic magmatism

The difficulties of relating magmatic activity of continental regions to the influence of long-lived stationary mantle plumes or hot spots were reviewed in Section 15.4.1 with respect to Younger Granites, and again in Section 16.2.2. Similar considerations apply to other regions of Cenozoic volcanism, in West Africa and elsewhere.

The basin-and-swell structure of the African continent began to take shape soon after the end of the Pan African event, as the continental crust and lithosphere adjusted to lateral inhomogeneities of structure and composition. Some parts became warmer and thicker and more buoyant than others, possibly in part because of regional variations in the concentration of radioactive heat-producing elements. The swells have been the site of uplift and potential or actual igneous activity at intervals throughout the succeeding several hundred million years. Their positions are fixed with respect to the lithospheric plate itself and must move with it over the underlying mantle.

So far as West Africa is concerned, for example,

the Hoggar and Aïr regions both experienced repeated uplift and doming over the last few hundred million years, and there has been igneous activity in the Palaeozoic, possibly in the Cretaceous, and again in the Cenozoic. The Jos Plateau was a site of major igneous activity for much of the Jurassic and became a volcanic area again during the Tertiary. The sedimentary basins have a similarly prolonged and varied history. The Taoudeni and Volta Basins have existed for upwards of 1000 Ma, and they are still relatively low-lying areas, as is the rather younger Iullmedden Basin, whose history nonetheless extends back to the Lower Palaeozoic; while the Chad Basin, younger still, appears to have been subsiding and accumulating sediments from the Cretaceous to the present time.

These features have retained their individual identities relative to one another as the position of the African continent has changed in response to plate movements. The basic pattern has not altered, even though the continent as a whole has undergone several cycles of uplift since it became a coherent shield area at the end of the Pan African. Within these uplift cycles there were almost certainly differential vertical movements, at times diminishing the contrast between basins and swells, at times accentuating it.

There are two ways in which regional uplift on a continental scale can happen, and they need not be mutually exclusive. Continental lithosphere is thicker than oceanic lithosphere and has an insulating effect on the underlying mantle if it is stationary for a significant length of time (100 Ma or more). The mantle will become warmer, less dense and more buoyant, and uplift will result. Alternatively, plate movements can bring continental lithosphere over areas of mantle that are already relatively warm, and once again uplift will result. In either case, the uplift can occur by phase changes in the mantle, e.g. from eclogite to granulite, or by partial melting in the mantle and the base of the crust, or both. These are very large-scale processes, however, and they do not significantly affect the distribution of the basin-and-swell features.

Over the African continent as a whole, swells are developed on crust of all ages, from Archaean to Pan African, though most of the major swells appear to be in Pan African terranes. Cenozoic igneous activity is virtually confined to these non-cratonic areas, and the same applies to a good deal of the Mesozoic magmatism as well. This is scarcely surprising, for older continental crust is likely to be cooler and more refractory than younger crust. It has been relatively depleted of radioactive heat-producing elements, partly because of prolonged erosion, but chiefly because of straightforward radioactive decay over longer periods of time.

The controls on the development of basins and swells and magmatic activity in continental areas would seem to reside mainly in the lithosphere itself, rather than in the underlying mantle. There is evidence that igneous activity away from plate margins is more likely to break out on a large scale when lithospheric plates are slow moving or stationary. But it will be inhomogeneities or variations in the properties of the lithospheric plates themselves that will primarily determine where and when thermal instabilities in the underlying mantle lead to igneous activity in the crust.

16.4 Diamond fields in West Africa

The most important diamond area in West Africa is still the Birim field of Ghana, with annual production of 1–1.5 million carats (Sec. 4.7.3). But these diamonds come from Proterozoic sediments, as do some of those mined in Ivory Coast. Otherwise, the diamond fields of West Africa (Fig. 16.2) are related to the kimberlite pipes which appear to have been emplaced at various times up to the Cretaceous (Sec. 16.1.1). Most of these diamond fields are alluvial, and though the primary kimberlite source of some of them is known (e.g. in Sierra Leone), in other places the source has not yet been found. Locating the primary sources of alluvial diamonds is important for two reasons. First, it helps geologists to predict where further alluvial deposits may be discovered, and secondly, as the alluvial fields become exhausted, diamonds can be won from the host kimberlite, though this is less attractive economically.

Sierra Leone rivals Ghana in diamond production, for annual exports were over two million carats in the early 1970s. This total fell to around 700 000 carats in the succeeding decade, but rises in the price of diamonds over the same period meant an increase of around 60% in the actual export value. The decline in production is largely attributed to exhaustion of the more easily accessible and profitable deposits in the Yengema and Tongo areas. Less easily exploited deposits require greater capital expenditure on mining equipment, particularly if the kimberlite pipes themselves are to be worked

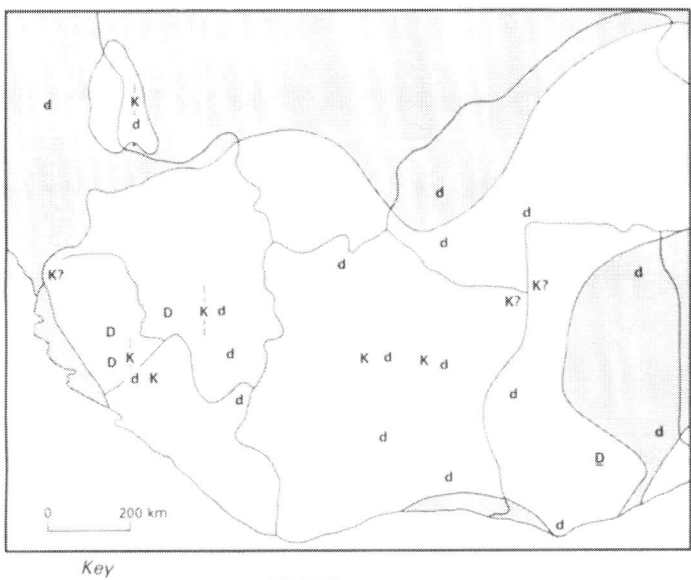

Figure 16.2 Some kimberlites (K) and diamond fields of major (D) and minor (d) importance in West Africa, in relation to cratonic basement and overlying sediments. Most of the diamonds are believed to be derived from kimberlites which have yet to be discovered. The Birim field of southern Ghana (D) is from Birimian metasediments.

Key

☐ cratonic basement   ☐ overlying sediments

directly, which has already commenced in a few places. There is also the perennial problem, common to all diamond mining ventures anywhere in the world, of illicit mining and smuggling of the stones.

In Liberia, diamonds account for some 7% of export revenue. Production is mainly from alluvial fields that extend across the border from the rich deposits in Sierra Leone. Annual output has been estimated to be of the order of 300 000 carats, worth over 30 million dollars, in the 1970s, but production has fallen and accurate estimates are no longer possible because of an increase in smuggling.

In Guinea, diamonds have been mined since at least 1936, but annual production appears never to have exceeded 100 000 carats. However, reserves are enormous, the most promising region being in the south-west of the country, around Kissidougou, which is not far from the Sierra Leone border. Reserves are estimated to be $300–400 \times 10^6$ carats, worth around 50 billion dollars at prevailing prices. Commercial production was planned to reach 250 000 carats annually in 1984.

There are two confirmed kimberlite fields in Ivory Coast, at Seguela and Kanangone near the centre of the country, and, though there are alluvial diamond deposits nearby, it is not clear which of them are derived from ancient Birimian rocks and which from younger kimberlites. Production of diamonds

from the Tortiya field recommenced in 1979, with an output of 50 000 carats, which is projected to rise in future.

The potential for diamond mining in countries on the West African craton is still considerable. The deposits in southern Mali (the Kenieba inlier, Fig. 16.2) appear to be considerable, and there are recorded discoveries in adjacent northern Guinea, as well as in northeastern Liberia, northern Ghana and southern Upper Volta.

16.5 Economic potential of Cenozoic volcanics

Apart from their uses as *crushed rock* for aggregates and road building, basaltic and related volcanic rocks are generally of little economic value as such. Extensive lava plateaux may have pyroclastic layers and well jointed flows that will provide small but useful *water*-bearing layers (aquifers, Part IV). Under the right conditions, basaltic rocks weather to kaolinite-rich *clays*. For example, among the Fluvio-Volcanics of the Jos Plateau there are clays that have some potential for ceramics and brick manufacture. Deep weathering of basaltic volcanics in the Aadmaoua region of Cameroun has produced a *bauxite* deposit estimated to contain over a billion tonnes of ore.

Rather fortuitously, basal gravel and sand layers of the Fluvio-Volcanics are also repositories of alluvial *cassiterite* derived from early unroofing of the Younger Granites, and some of these deposits are among the richest to be worked. Elsewhere on the Jos Plateau, the thin basalt flows of the youngest volcanoes are a hindrance to tin mining, for they occupy old river valleys and thus cover the alluvium with a hard resistant capping. This interferes with geophysical prospecting for concentrations of cassiterite and columbite in the underlying alluvium, and it makes exploratory drilling very expensive. Potentially rich deposits of ore have been effectively sterilised by these basaltic caps, but they will eventually be found and worked as the more easily accessible deposits become exhausted.

River gravels south of the Jos Plateau contain alluvial concentrations of megacryst corundum and zircon, weathered out of basaltic lava flows (Sec. 16.2.2). Some of the corundum crystals are gem quality *sapphires*, and the larger zircon crystals could have some potential as semi-precious stones. Indeed, the alluvial deposits of the Jos Plateau region are now known to contain gem quality crystals derived from basement pegmatites (Sec. 6.5). Younger Granites (Sec. 15.5) and Cenozoic volcanics. They may in future become more important as a source of gemstones than as a source of tin and niobium–tantalum ores.

Bibliography

Anderson, D. L. 1982. Hotspots, polar wander, Mesozoic convection and the geoid. *Nature* 297, 391–3.

Bardet, M. 1973, 1974, 1977. *Géologie du diamont*. Mem. Bur. Rech. Geol. Min. 83, Nos 1, 2 and 3.

Black, R. and M. Girod 1970. Late Palaeozoic to Recent igneous activity in West Africa and its relationship to basement structure. In *African magmatism and tectonics*, T. N. Clifford and I. G. Gass (eds). Edinburgh: Oliver & Boyd.

Briden, J. C. and I. G. Gass 1974. Plate movement and continental magmatism. *Nature* 248, 650–3.

Culver, S. J. and H. R. Williams 1979. Late Precambrian and Phanerozoic geology of Sierra Leone. *J. Geol. Soc. Lond.* 136, 605–18.

Dautria, J. M., J. Girod and O. Rahaman 1984. The upper mantle beneath eastern Nigeria: inferences from ultramafic xenoliths in Jos and Biu volcanics. *J. Afr. Earth Sci.* 2, 331–8.

Dawson, J. B. 1970. The structural setting of African kimberlite magmatism. In *African magmatism and tectonics*, T. N. Clifford and I. G. Gass (eds). Edinburgh: Oliver & Boyd.

Dunlop, H. M. and J. G. Fitton 1979. A K–Ar and Sr-isotopic study of the volcanic rocks of the island of Principe, West Africa – evidence for mantle heterogeneity beneath the Gulf of Guinea. *Contrib. Mineral. Petrol.* 71, 125–31.

Fabre, J. 1976. *Introduction à la géologie du Sahara Algérien*. Algiers: SNED.

Fitton, J. G. 1980. The Benue Trough and Cameroun line – a migrating rift system in West Africa. *Earth Planet. Sci. Lett.* 51, 132–8.

Fitton, J. G., C. R. J. Kilburn, M. F. Thirlwall and D. J. Hughes 1982. 1982 eruption of Mount Cameroon, West Africa. *Nature* 306, 327–32.

Furon, R. 1963. *Geology of Africa* (English edn). London: Oliver & Boyd.

Gass, I. G., D. S. Chapman, H. N. Pollack and R. S. Thorpe 1978. Geological and geophysical parameters of mid-plate volcanism. *Phil. Trans. R. Soc. Lond. A*, 581–97.

Girod, M., J. M. Dautria and R. de Giovanni 1981. A first insight into the constitution of the upper mantle under the Hoggar area (southern Algeria): the lherzolite xenoliths in the alkali-basalts. *Contrib. Mineral. Petrol.* 77, 66–73.

Grant, N. K, D. C. Rex and S. J. Freeth 1972. Potassium–argon ages and strontium isotope ratio measurements from volcanic rocks in northeastern Nigeria. *Contrib. Mineral. Petrol.* 35, 277–92.

Haggerty, S. E. 1982. Kimberlites in western Liberia: an overview of the geological setting in a plate tectonic framework. *J. Geophys. Res.* 87, 10, 811–26.

Hastings, D. A. 1974. Proposed origin for Guianian diamonds. *Geology* 2, 475–6.

Hastings, D. A. and W. E. Sharp 1979. An alternative hypothesis for the origin of West African kimberlites. *Nature* 277, 152–3.

Irving, A. J. and R. C. Price 1981. Geochemistry and evolution of lherzolite-bearing phonolitic lavas from Nigeria, Australia, East Germany and New Zealand. *Geochim. Cosmochim. Acta* 45, 1309–21.

Le Bas, M. J. 1971. Per-alkaline volcanism, crustal swelling and rifting. *Nature, Phys. Sci.* 230, 85–7.

Liotard, J. M., C. Dupuy, J. Dostal and G. Cornen 1982. Geochemistry of the volcanic island of Annobon, Gulf of Guinea. *Chem. Geol.* 35, 115–28.

Olade, M. A. 1978. Early Cretaceous basalt volcanism and initial continental rifting in Benue Trough, Nigeria. *Nature* 273, 458–9.

Quin, J. P. and P. Fraudet 1976. Sur la nature du volcanism de la région de Thies, Senegal. In *African Geology*, Tsegaye Hailu (ed.). Ibadan: Geol. Soc. Africa.

Smith, A. G. 1982. Late Cenozoic uplift of stable continents in a reference frame fixed to South America. *Nature* 296, 400–4.

Spengler, A. de, J. Castelain, J. Cauvin and M. Leroy 1966. Le bassin Secondaire-Tertiaire du Sénégal. In *Sedimentary basins of the African coast*, D. Reyre (ed.). Paris: ASGA.

Stillman, C. J., H. Furnes, M. J. Le Bas, A. H. F. Robertson and J. Zielonka 1982. The geological history of Maio, Cape Verde Islands. *J. Geol. Soc. Lond.* **139**, 347–62.

Thorpe, R. S. and K. Smith 1974. Distribution of Cenozoic volcanism in Africa. *Earth Planet. Sci. Lett.* **22**, 91–5.

Turner, D. C. 1978. Volcanoes of the Biu Plateau, north-eastern Nigeria. *J. Mining Geol.* **15**, 47–64.

Turner, D. C. and P. Bowden 1979. The Ningi-Burra complex, Nigeria: dissected calderas and migrating magmatic centres. *J. Geol. Soc. Lond.* **136**, 105–20.

Umeiji, A. C. and M. Caen-Vachette 1983. Rb–Sr isochron for Gboko and Ikyuen rhyolite and its implication for the age and evolution of the Benue Trough, Nigeria. *Geol. Mag.*, **120**, 529–33.

Verheijen, P. J. T. and D. E. Ajakaiye 1979. Geophysical anomalies over a pipe suspectedly kimberlite in the Precambrian metamorphic schist belt of northern Nigeria. *Tectonophysics* **17**, 293–303.

Vogt, P. G. 1981. On the applicability of thermal conduction models to mid-plate volcanism: comments on a paper by Gass *et al. J. Geophys. Res.* **86**, 950–60 (reply, *ibid.* 961–6).

Williams, H. R. and R. A. Williams 1977. Kimberlites and plate-tectonics in West Africa. *Nature* **270**, 507–8.

Wright, J. B. 1976. Volcanic rocks in Nigeria. In *Geology of Nigeria*, C. A. Kogbe (ed.). Lagos: Elizabethan Press.

Part IV

THE QUATERNARY OF WEST AFRICA

17 Introduction: earthquakes, volcanoes and meteorites

SUMMARY

The Quaternary should not be viewed in isolation but as a link with the more distant geological past. In many places, the present cycle of weathering and erosion began long ago, and is responsible for some important economic mineral deposits. The tectonic and magmatic effects of continued seafloor spreading and regional uplift are still felt as occasional earthquakes, and they have provided a number of mildly active and dormant volcanoes. A geological event without known precedent in this region is the meteorite impact that produced the crater now occupied by Lake Bosumtwi in Southern Ghana, and the tektites of the eastern Ivory Coast.

17.1 Introduction

In the final Sections of this text a number of Quaternary processes will be considered, partly to show that geological processes initiated millions or even tens of millions of years ago are still in progress today. In West Africa these processes have been responsible for much of the economic mineral wealth of the region, especially the formation of residual deposits of bauxite, the secondary enrichment of manganese and iron ores, and the concentration of alluvial diamonds, gold, rutile, cassiterite and other minerals into placer deposits. Some of the superficial sediments have uses in their own right, especially laterite and alluvial clays, but the most important resource of Quaternary age is water, without which no society can function.

The present landscape of West Africa has been shaped by a combination of climatic influences and tectonic forces that were active in Tertiary times if not before, and continued to be active through the Quaternary. This chapter will concentrate on features that are primarily of tectonic origin, the next will deal mainly with surface processes and economic aspects.

17.2 Earthquakes and volcanism

West Africa is generally considered to be an aseismic region, that is, one characterised by low earthquake risk. That is because of its long distance from major plate margins, especially subduction zones. Earth-quakes are relatively common and severe to the north and east of the continent. They also occur in West Africa, however, as a result of adjustments along the oceanic fracture zones where they intersect the continental margin (Fig. 11.8). Major earthquakes affected the region round Accra in 1862, 1906 and 1939, and there was seismic activity near the Volta Dam in 1970. These events are believed to have resulted from movements along the NNE trending Akwapim Fault system, which shows strong evidence of activity during the Quaternary, and which projects seaward into the Romanche fracture zone. Records of the major earthquakes have been studied in some detail, with a view to estimating potential hazards for future major engineering works in this part of southern Ghana. There has been intermittent seismic activity on the Cameroun coast, in the Kribi area, south of the Cameroun line, where oceanic fracture zones intersect the continental margin. Other faults related to the coastal sedimentary basins may have been active during the Quaternary. It is more difficult to explain the 6.2 magnitude Guinea earthquake of December 1983, centred close to 12°N, 14°W, that is well within the continental part of this supposed aseismic region. Over 400 people died, thousands were made homeless and the effects were felt throughout Guinea and Sierra Leone, Senegal, Gambia and Guinea–Bissau.

Earthquakes are also associated with volcanic activity. Mount Cameroun is still active and studies of magma volumes erupted at intervals since the early twentieth century suggest that the next significant eruption may occur between AD2003 and 2010;

though this does not preclude the possibility of smaller eruptions before then.

Many of the basaltic cones elsewhere in the Gulf of Guinea province have a very youthful aspect and radiometric dates confirm that they were active until very late in the Quaternary (Sec. 16.2.2). They appear to represent waning stages of volcanic activity, however, and further outbreaks in these areas would seem to be unlikely.

17.3 Lake Bosumtwi

Lake Bosumtwi lies about 30 km south-east of Kumasi in southern Ghana and is now generally recognised to be a meteorite impact crater. The lake is roughly circular in outline, nearly 8 km in diameter, and its surface is about 100 m above sea level. It is surrounded by a steep-sided crater that has been excavated in Lower Birimian metasediments with their characteristic steep dips and regional NE–SW trend. The crater rim is between 10 and 11 km in diameter and it stands between about 150 and 500 m above the level of the lake. The inner walls of the crater are steep-sided, with slopes of up to 30 or 40°, becoming less steep towards the lake, which has a gently sloping bottom, the greatest depth being about 80 m.

The outer slopes of the crater rim are less steep than the inner slopes and are surrounded by a ring-shaped depression between 1 and 5 km in width. Beyond it is a ridge about 30 to 80 m above the level of the depression. There is an escarpment on the outer side of this ridge that maintains a fairly uniform distance of about 10.5 km from the centre of the lake.

The lake is a basin of internal drainage, with a radial pattern of streams flowing into it. For many years, the Bosumtwi crater was believed to be of volcanic origin, though formation by meteorite impact was first suggested as long ago as 1931. Geophysical investigations have yielded inconclusive results so far, and the evidence that Bosumtwi is a meteorite crater comes partly from the physiographical relationships just described, but mainly from the rocks that outcrop in and around the depression.

17.3.1 The evidence for an impact origin
Shattering and melting of the rocks at sites of meteorite impact produce rock types that are superficially similar to the pyroclastic products of explosive volcanism. Closer examination reveals some distinctive features, however.

Breccias outcrop abundantly around Lake Bosumtwi, consisting of two main kinds, which occur in different parts of the area. From the lake side as far as the outer rim of hills are found coarse breccias in a fragmentary matrix. They typically consist of only one rock type, which is normally phyllite, and shattered phyllite can be seen grading into unshattered phyllite or interbedded with it. These breccias are interpreted as resulting from fragmentation of the rocks in place during the meteorite impact, the phyllites being most susceptible to this process.

The other kind of breccia is principally exposed on the outer rim of hills, where it outcrops along stream valleys. It consists of angular blocks of many different rock types, commonly a metre or so across, but sometimes up to 5 m. It rests unconformably upon the Birimian rocks, for subhorizontal bedding is sometimes visible and the grain size becomes noticeably coarser towards the lake. The thickness of the deposit is at least 15 m, and it is interpreted as a superficial blanket formed by ejection of fragments during formation of the crater.

Impactite is formed by melting of rocks during meteorite impact, and has been found in two places at Bosumtwi. In the depression north of the crater it outcrops for about 1 km, overlying Birimian rocks in a stream section, and in places is up to 13 m thick. It is yellowish grey with an earthy matrix containing occasional rock fragments and variable amounts of pumice in irregularly shaped lumps up to 15 cm across. The other outcrop is exposed along streams outside the rim south of the lake, where the rock consists of up to 40% of lumps of dark vesicular glass, along with angular fragments of Birimian rocks.

Analyses of glass and pumice from the Bosumtwi crater suggest compositions approximating to andesite, which is understandable, as they represent melted metasedimentary rocks of probably original volcanoclastic origin. Chemically, therefore, these materials are not significantly different from some terrestrial volcanic rocks, and were originally interpreted as volcanic agglomerates. However, detailed petrographic studies have revealed their impact melt origin. **Coesite** is a high-pressure polymorph of silica found only at meteorite craters and it has been identified in glass from the southern outcrop. Metallic spherules a few microns across have been found in the glass and contain up to 5% of Ni, also consistent with a meteorite origin. In addition, typically refractory minerals such as ilmenite and zircon in Birimian rocks have been melted or decomposed, effects which require temperatures far above those that occur during volcanic activity and have only been encountered at meteorite craters.

The age of the crater has been found by dating the impactite glasses. Several measurements using K/Ar and fission track methods have yielded an average age close to 1.3 Ma. The Rb/Sr isochron age of the glasses is about 2000 Ma, similar to that of the Birimian country rocks round the crater, thus further confirming its impact origin.

17.3.2 Ivory Coast tektites

Tektites are small glass bodies, rarely more than a few centimetres across, with a variety of shapes ranging from spherical to spindle shaped to dumbbell shaped. Originally thought to be of extraterrestrial origin, they are now generally accepted as having been formed by the ejection of drops of molten rock during meteorite impacts, a conclusion strengthened by the fact that they are found in the vicinity of meteorite craters. The Ivory Coast tektites are found around Ouellé, about 250 km west of Lake Bosumtwi. Geochemically and isotopically they are very similar to the crater glasses and the Birimian rocks at Bosumtwi, and they give virtually identical K/Ar, fission track and Rb/Sr isochron ages.

17.3.3 Lake sediments

Within the lake and in the lower courses of streams flowing into it are finely laminated silts and carbonaceous clays containing leaf impressions and occasional fossil fish. In places the succession is interrupted by lenses of coarse conglomerate and gravel, and the proportion of these increases away from the lake until they entirely replace the silts and clays. There are also small outcrops of cross-bedded sands and gravels with gastropod shells and fish remains.

The silts are thought to be the products of density currents flowing into the lake, perhaps resulting from rain storms on the crater walls, while the clays represent normal very slow deposition in deep water. The conglomerates may be brecciated rocks of the crater wall, resedimented as a result of landslides on the slopes of the crater, and the cross-bedded sands and gravels were perhaps deposited in small stream deltas.

A clearly defined terrace between 50 and 100 m wide encircles the lake at an elevation of about 125 m above the water level. It merges with the lower slopes of a valley which cuts through the eastern rim of the crater and which functioned as an overflow channel in the past, so that the terrace represents the highest possible level of the lake.

Field investigations supported by radiocarbon dating of lake sediments suggest that the lake was at a low level for much of the late Pleistocene. It began to rise about 13 000 years ago, and thereafter periods of generally high water level alternated with marked regressions at about 10 000, 8000, 4000 and 1000 years ago. It appears that the level has been rising since then, though possibly with minor fluctuations; in the mid-19th century it was about 30 m lower than at present.

Lake Bosumtwi is the only large natural lake in the West African forest zone and, having a small internal drainage system, it is uniquely placed for recording the climatic fluctuations in this zone during the Quaternary.

Bibliography

Bacon, M. and J. K. A. Banson 1979. Recent seismicity of southeastern Ghana. *Earth Planet. Sci. Lett.* **44**, 43–6.

Blundell, D. J. 1976. Active faults in West Africa. *Earth Planet. Sci. Lett.* **31**, 287–90.

Burke, K. 1969. Seismic areas of the Guinea coast where Atlantic fracture zones reach Africa. *Nature* **222**, 655–7.

Burke, K. 1971. Recent faulting near the Volta Dam. *Nature* **231**, 439–40.

Fitton, J. G., C. R. J. Kilburn, M. F. Thirlwall, and D. J. Hughes 1982. 1982 eruption of Mount Cameroon, West Africa. *Nature* **306**, 327–32.

Jones, W. B., M. Bacon and D. A. Hastings 1981. The Lake Bosumtwi impact crater, Ghana. *Geol. Soc. Am. Bull.* **92**, 342–9.

Quaah, A. O. 1982. A study of past major earthquakes in southern Ghana using intensity data. *Tectonophysics* **88**, 175–88.

Shaw, H. F. and G. J. Wasserburg 1982. Age and provenance of the target materials for tektites and possible impactites as inferred from Sm–Nd and Rb–Sr systematics. *Earth Planet. Sci. Lett.* **60**, 155–77.

Smit, A. F. J. 1962. The origin of Lake Bosumtwi and some other problematic structures. *Ghana J. Sci.* **2** (2), 176–96.

Talbot, M. R. and G. Delibrias 1980. A new Pleistocene-Holocene water-level curve for Lake Bosumtwi, Ghana. *Earth Planet. Sci. Lett.* **47**, 336–44.

Thompson, T. F. 1970, 1973. Holocene tectonic activity in West Africa dated by archaeologic methods. *Geol. Soc. Am. Bull.* **81**, 3759–64, **83**, 1197–9.

18 *Geomorphology, Quaternary deposits and water resources*

SUMMARY

Much of the southern part of West Africa is relatively low lying and of low relief. It includes the drought-prone Sahelian belt south of the Sahara. The drainage pattern is naturally westward and southward towards the sea, but there is evidence that some major rivers originally flowed into the interior, before being captured and diverted to the south. The relatively low salinity of Lake Chad is controlled partly by infiltration, partly by precipitation of mineral phases.

A number of erosion surfaces have been identified and correlated with cycles of regional uplift during the Tertiary and Quaternary, and with changing sea levels later in the Quaternary. Alluvial sediments contain placer deposits of gold and diamonds and other minerals.

Laterites are extensively developed. Primary laterites formed *in situ* on older erosion surfaces probably during the Tertiary. This lateritisation phase may have been responsible also for the formation of residual bauxite and manganese ores, and for the enrichment of iron ores. Secondary laterites resulted from erosion of the older laterites. Quaternary deposits include aeolian drift and coastal sediments.

Some of the superficial deposits are used for constructional purposes, especially the laterites. Because of the great extent of these deposits, they present problems both for civil engineering and mineral exploration projects. Storage of surface and near-surface water is improved by dam construction. There is a good deal of ground water in fracture systems among basement rocks, and sandstones and limestones among the sediments also provide good aquifers. There is artesian water in the larger sedimentary basins.

18.1 Geomorphology

The present-day geomorphology of any region is of course a culmination of events that have shaped the landscape over a long period of time, not merely the last million or so years of the Quaternary. This applies particularly to tropical and equatorial regions remote from the Pleistocene ice caps of the higher latitudes. In the northern glaciated areas, the superficial deposits are mainly of late Quaternary age, because glaciers and melt waters eroded and redistributed older deposits.

The main physiographic features of West Africa soon become apparent on inspection of any up-to-date atlas. Most of the region lies below 500 m in altitude. Higher ground in western parts of the craton, in southern and southwestern parts of the Taoudeni Basin, and in the eastern Pan African domain, is mostly between 500 and 1000 m altitude. Parts of the craton are above 1000 m, but there are more extensive highlands in the eastern Pan African domain: the Jos and Mambilla Plateaux, the Cameroun Mountains and, further north, the Hoggar and Aïr massifs. In these eastern areas Younger Granites and Cenozoic volcanoes are partly responsible for

peaks that exceed 3000 m in places, though the general level of the ground is generally not above 1500 m.

Most of the coastal belt is thickly vegetated with tropical rain forest, coastal swamps and lagoons, though there are locally belts of relatively dry grassland as well. The forest belt gives way northwards to progressively less densely wooded and more open country of grassland and savanna. North of about 13°N latitude and south of the Sahara Desert lies the semi-arid to arid **Sahelian belt**.

Over huge areas the relief is very low. In basement regions, the monotony of the plains is broken by inselbergs of granite, migmatite or gneiss, that rise up to 600 m above the general level of the surrounding country. Resistant supracrustals, especially quartzites, form ridges and ranges of hills. In regions underlain by sediments there are escarpments and flat-topped hills, though these nowhere approach the height of basement inselbergs. Upland areas have a more varied landscape, especially where there has been volcanic activity, and there are areas of moderate relief near some of the larger rivers and their tributaries.

154

18.1.1 Drainage

Most rivers in the southern part of West Africa flow towards the south or west, more or less directly into the Atlantic, with two notable exceptions. One is the internal drainage system of the Chad Basin, the other is the Niger River, the longest in the region. The Niger rises in the highlands of Guinea and flows north-east to near Timbuktu, before turning east and then south-east towards its confluence with the Benue, and finally south to its delta in the Gulf of Guinea. There are two main watersheds in the southern part of West Africa. One separates the Niger catchment from rivers flowing into the Gulf of Guinea, the other surrounds the Chad Basin (cf. Fig. 10.3).

Headward erosion by southward flowing river systems has resulted in the capture of rivers that formerly flowed towards the interior. For instance, the northeastward course of the upper Black Volta suggests that it was once part of the Niger system, but was captured by the southerly flowing lower Black Volta. This in turn once flowed directly into the sea but was itself captured by a tributary of the White Volta. Further east, in Nigeria, the Gongola River formerly flowed north-east from the Jos Plateau into the Chad Basin, until it was captured by a tributary of the Benue.

The Niger itself only became a single river in the Quaternary as a result of river capture. The upper Niger originally spread out and terminated as an inland delta in a large basin of internal drainage now represented by the seasonally fluctuating marshes and shallow lakes of the Macina, which lies some 300 km upstream from Timbuktu, about 300km north-east of Bamako (Fig. 9.1). The ancestral lower Niger rose in the Adrar des Iforas, and one of its tributaries eventually cut back to capture the inland lake and the upper Niger which flowed into it. The southern Taoudeni Basin is thus no longer a region of internal drainage.

As might be expected, sediments in the Macina resemble those of the Chad Formation, comprising clays, calcareous clays, sands and diatomite. They are referred to the Continental Terminal, however, and they do not exceed 100 m in thickness, so the Macina was not an area of major subsidence. The sediments are capped in several places by thick laterite, which is in turn overlain by alluvial deposits and dune sands.

The largest natural body of open water in West Africa is Lake Chad, which lies at about 275 m above sea level. It is supplied almost entirely by the peren-

nial River Chari, which rises in mountains to the south-east. The lake has no surface outflow, for the Chad Basin is one of internal drainage and it is extremely shallow, 3–4 m deep at most. Even in the mid-1960s, before the most recent Sahelian drought, the maximum depth was not more than about 7 m.

In view of the enormous evaporation that must occur, the average salinity of the Lake Chad waters is remarkably low, at around 400 mg l^{-1} of dissolved salts, compared with about 120 mg l^{-1} for average river water and 35 g l^{-1} for normal sea water. There is an overall increase in the salinity of Lake Chad from the south (near the Chari delta) to the north. The waters are of the ordinary Ca–Mg–Na–HCO$_3$ type that characterise most rivers, and the low salinities appear to result in part from infiltration, in part from precipitation of clay minerals of the montmorillonite (smectite) type. The somewhat higher salinities in northern parts lead to the precipitation of a variety of salts, including calcite and trona (a complex sodium carbonate mineral).

The northeastern shore of the lake is a vast archipelago of elongate islands and sand banks, the remains of a dune field that may be 40 000 years old and represents a former dry period (Sec. 10.3.2). The dunes are made of quartz sands, with clays and carbonate muds in the depressions between them. Megachad reached its maximum extent between about 9000 and 5000 years ago, when water depths over the present Lake Chad were about 50 m, but were much greater over the Bodele depression to the north-east, which is at a lower altitude. This great lake had a southern outlet, which drained into the Benue River, (Fig. 10.3).

Lake Chad and its predecessors must to some extent resemble the Quaternary lake of the Macina, where the Niger had its inland delta. If natural processes were allowed to run their course, it is possible that active headward erosion by tributaries of the Benue River would eventually result in the capture of the Lake Chad system. This would link the Chari and Benue into a single great river second only to the Niger in size and importance. It would be many thousands of years before that happened, however, unless it were seen to be a desirable project, in which case it could probably be engineered within a few years. Conversely, it can as easily be prevented from ever happening.

18.1.2 Erosion surfaces

Erosion surfaces have been identified and mapped in various parts of West Africa, but their number and the ages assigned to them differ from region to region. The development of these surfaces at successively lower levels is the result of intermittent uplift during Tertiary and Quaternary times. There has also been some crustal warping, however, and this presents problems when attempts at long-range correlation are made on the basis of accordant surface or summit levels.

For example, the Jos Plateau in Nigeria stands 1200–1300 m above sea level and it has been correlated with the late Jurassic **Gondwana Surface** recognised elsewhere in Africa. Numerous inselbergs of Jurassic Younger Granite rise above the general level of the plateau, however. If any remnants of the Gondwana Surface are preserved here, they would presumably be found among the summits of those inselbergs.

The Gondwana Surface must by definition predate the break-up of Gondwanaland, and it must form the unconformity between basement rocks and Cretaceous sediments beneath the younger basins. Gentle warping of this surface may have begun in the Cretaceous, initiating the differential uplifts of the Jos and Mambilla Plateaux and the Hoggar, Aïr and Tibesti massifs, which provided the marginal swells to these basins. These regions are highlands at the present time, but during the Cretaceous the whole of West Africa must have been much closer to sea level, with low-lying land areas of subdued relief in between the basins (except near the Benue Trough).

Even when the seas retreated early in the Tertiary, West Africa remained near sea level for some time, a vast **peneplain** extending across both basement and sediments. This would represent the post-Gondwana or **African Surface** recognised in other parts of the continent and variously dated as late Cretaceous to mid-Tertiary in age. Because the whole West African region was near to sea level during most of the Upper Cretaceous and early Tertiary, there can have been little erosion, and where the African Surface developed on older rocks it could virtually coincide with the older Gondwana Surface over large areas.

Major uplift and warping began later in the Tertiary, with the onset of Cenozoic volcanism during the Miocene. The various cycles of planation and formation of **pediments** identified in different parts of West Africa can be referred to this period. Remnants of the African Surface have been recognised between about 500 and 800 m in western parts of the craton, and also in northern Nigeria. Here the surface slopes down gently northwestwards to the

Iullmedden Basin in the north-west and up towards the Jos Plateau, where younger pediment surfaces have been cut into it. The older surface must also have an overall seaward slope, because near the coast it approximates to the surface of the sediments in the coastal basins. Some planation surfaces found at levels lower than about 500 m may therefore also be remnants of the post-Gondwana (African) surface.

One of the most obvious examples of crustal warping is the gap between the Bida and Sokoto Basins (Fig. 11.1). The sediments in these basins were once continuous until a broad northeasterly axis of uplift caused the Niger River to incise its channel through the sediments and into the underlying basement.

Alluvial silts and sands and occasional gravels are found along virtually all stream or river courses of any size. Natural terraces in these deposits provide evidence of rejuvenation and downcutting in response to continued uplift and changing sea level during the Pleistocene. Many of the placer concentrations of valuable minerals (gold, diamonds, cassiterite) are found in the coarser sands of streams and rivers rejuvenated during the Quaternary. The continuing rejuvenation also contributes to gully erosion, which affects many regions where there are soft sediments or thick superficial deposits on deeply weathered bedrock. Both gully and sheetwash erosion have been aggravated by overgrazing and clearance of bush for cultivation, leading to problems of soil conservation in parts of West Africa.

On a larger scale, western and northern Africa belong to what has been called 'low' Africa, which lies mostly below 1000 m, whereas southern and eastern regions comprise 'high' Africa, mostly above 1000 m (Fig. 18.1). The differential uplifts responsible for this broad subdivision of the continent are believed to be of late Cenozoic age, and a number of explanations for it have been advanced. Geophysical investigations indicate that the lithosphere is thinner below low Africa than below high Africa and it may be significant that the boundary between the two more or less coincides with a major negative gravity anomaly.

18.2 Laterites

The widespread development of laterites in West Africa is of limited assistance in the identification and correlation of erosion surfaces. Laterite formation takes a long time and it will begin wherever the

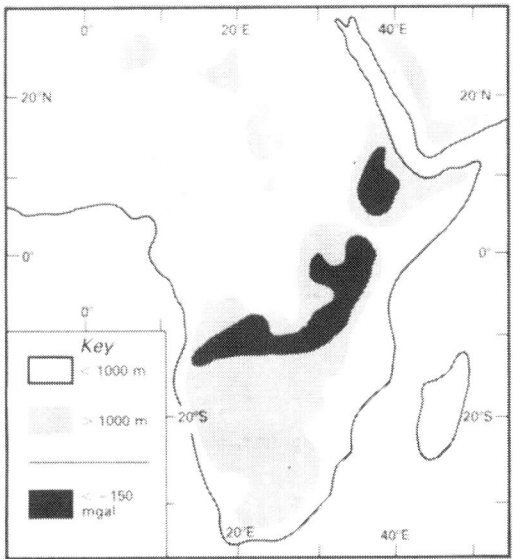

Figure 18.1 The division of 'high' and 'low' Africa in relation to the belt of maximum negative Bouguer gravity anomaly. The gravity anomaly under most of the rest of 'high' Africa is −125 (±25) mgal, beneath much of 'low' Africa it is −50 (±30) mgal, but the higher areas in this region are close to or roughly coincident with areas of −90 (±10) mgal.

conditions are favourable. Lateritic deposits of identical appearance are therefore not necessarily of the same age.

Laterite is reddish to brownish in colour and it may be vesicular, concretionary, cellular, vermicular, slag-like, pisolitic, concrete-like, or gravelly in texture. It consists chiefly of ferric oxides with minor amounts of alumina and manganese oxides and variable quantities of quartz grains and fragments, as well as a substantial proportion of clay minerals. The hardness is variable, but laterite usually shatters easily when hammered. Bauxite resembles laterite, the chief difference being that alumina predominates greatly over iron oxides. Superficial enrichment of manganese ores in the weathering zone leads to the formation of laterite-like deposits, but here of course manganese is greatly enriched at the expense of iron and alumina.

Both primary and secondary laterites are recognised. Primary laterites form *in situ* by iron enrichment in the soil profile, generally through the removal of other constituents in solution. Laterites therefore develop more rapidly upon iron-rich parent materials such as basalts, but thick laterites are widespread in basement areas where they have often

formed on iron-deficient gneisses. There is normally an unbroken transition from weathered bedrock into laterite, and in basement areas steeply inclined quartz veins can be traced up into the laterite layer without noticeable displacement. In some places, the continuity is broken, however, and there is a layer of angular or rounded quartz pebbles just above the weathered bedrock, interpreted as the remains of sheetwash deposits laid down on the peneplain surface before lateritisation commenced. Laterite formation is favoured by low relief, poor drainage, a warm climate with alternating wet and dry seasons, and a fluctuating water table near the surface, which enables the insoluble ferric oxides to be precipitated in the zone of aeration.

In many parts of West Africa there are flat-topped erosion residuals of **high-level laterite**, also called peneplain laterite. They are rarely more than 10 to 15 m high, but they have steep slopes and in places are almost devoid of vegetation, with hard sterile surfaces. They occur on both basement and sediments and are considered to be the dissected remains of a once virtually continuous sheet that covered much of West Africa and extended as far north as about 15°N.

The formation of these laterites on basement and older sediments could well have commenced in the Cretaceous and extended to the younger sediments during Lower to Mid-Tertiary times, when West Africa as a whole probably still lay near to sea level. The main formation of bauxite and manganese ores and the secondary enrichment of iron ores would have occurred at the same time. There is no positive evidence to support these inferences, however, and it must be remembered that in many places the laterites are part of the Continental Terminal which comprises mainly fluviatile and lacustrine sediments whose age may be anything from Eocene to Pliocene (Sec. 10.2.2). All that can be said with confidence about the age of the laterites is that in a few places they are demonstrably Pliocene or older, because they are overlain by, for example, Chad Formation sediments and Quaternary basalts (in northeastern Nigeria), or by Quaternary alluvium of the Niger (in southern Mali).

Primary laterites are also developed on the later pediment surfaces cut into the ancient peneplain during successive phases of Cenozoic uplift. They are generally thinner than the high-level laterites, but the distinction between the two cannot always be made with confidence and this can make the correlation of erosion surfaces more difficult.

Secondary laterites are generally quite easy to identify on account of their generally detrital or gravelly aspect. They mantle the slopes of the high-level laterite plateaux and form a major component of the gravels that have been spread across many of the lower pediment surfaces. Detrital secondary laterites can form quite thick spreads of gravel, often mixed with quartz fragments and frequently cemented into a compact mass by iron oxides. Detrital laterites are also forming at the present time and cementation can be seen to be actively in progress in smaller seasonal stream channels.

Along with soil and vegetation and alluvial deposits, laterite is responsible for obscuring most of the bedrock in West Africa, as it is in many other tropical regions.

18.3 Other deposits

Coastal sediments in the Bullom Group of Sierra Leone range up to the Quaternary (Sec. 11.3) and the Quaternary sediments of the Chad Formation were described in Section 10.3.2. Mention was also made there of the beach ridges that mark former shorelines when Lake Chad was larger. The lake lies in the Sahelian belt, where wind-blown accumulations are a significant component of the superficial deposits. They include both fossil sand dunes and dune fields and the **aeolian drift**, a loess-like dust deposit brought in during each dry season on north-easterly winds (the Harmattan) from the Sahara. The dust consists mainly of clay and silt-sized particles of kaolinite and iron-stained quartz, along with plant pollen and spores and diatomite tests. The ENE–WSW alignment of dune ridges in some places, for example in the southwestern Chad Basin in northern Nigeria (where they show up well on satellite imagery), suggests that the pattern of prevailing dry season northeasterly winds may have persisted in West Africa for most of Quaternary times at least.

Studies of the distribution and stratigraphy of dune and drift deposits have helped to elucidate Quaternary climatic changes in West Africa. Arid and humid phases have been correlated respectively with glacial and interglacial periods in more northern latitudes. Quaternary sediments were also deposited in deltas and estuaries in the coastal belt, where longshore drift has formed barrier beaches and led to the development of lagoons. The extensive coastal sediments of Sierra Leone are in part of Quaternary

age and were described in Section 11.3. Well preserved raised beaches in Ghana are the result partly of continued uplift and partly of global sea level changes. The rise in sea level during the last 10000 years has drowned the mouths of many river valleys. Additional information on climatic changes has been obtained from examination of accumulations of wind-borne dust blown out to sea and deposited in the eastern Atlantic.

Of more immediate importance than climatic fluctuations on timescales of many thousands of years are the superimposed cycles of rainfall and drought that occur on timescales of decades and particularly affect the Sahelian belt. These cycles are recorded, for example, in fluctuation of run-off in major river systems, and in oscillations of the level of Lake Chad (there is no clear correlation between lake levels of Bosumtwi and Chad because they are in different climatic regimes). The 1970s was a period of severe drought in the Sahelian belt, and it persisted into the 1980s, but there is some evidence that wetter conditions might return by the 1990s.

18.4 Some economic aspects

As already mentioned, many of the mineral deposits described in earlier chapters are actually late Tertiary to Quaternary in age. This applies to the alluvial placers along river valleys, to the 'lateritic' ores of bauxite and manganese, and to the coastal deposits of rutile and lignite in Sierra Leone and of rutile further north along the coast of Gambia and Senegal. There are other heavy mineral concentrations in both river and beach sands in parts of West Africa, notably of ilmenite and zircon as well as rutile, but so far no additional economically workable deposits have been found. In Ghana, beach sands include economic concentrations of high-grade silica (glass) sands, and there are probably deposits elsewhere too.

Laterite itself is a well known road-building material in West Africa as in other parts of the continent, and in rural areas it is still extensively used for house building. Laterite developed on basement rocks is usually best for roads, as laterites from sandy sediments have too little clay, those from marine sediments usually too much. Ideally, the material should be well graded, with less than 10% clay-sized particles.

There is a general scarcity of gravels among the alluvial deposits of rivers, nothing to compare with the great fluvioglacial accumulations found along large rivers in higher latitudes. This is not a major problem, as virtually unlimited supplies of crushed rock aggregate for the construction industry can be obtained from quarries in basement rocks. Alluvial sands and clays along some streams and river valleys and along the coast line are also used for building, and in many places the clays are of suitable quality for making bricks and pottery. Indeed the vast quantities of such clays are a major resource in many parts of West Africa. They could be used much more than they are at present to help overcome shortages of indigenous sources of cement for construction purposes.

Over large areas, exploration for mineral deposits beneath the superficial cover is difficult because the cover is so thick and extensive and obscures so much of the bedrock geology. Mapping and mineral exploration can be supplemented by geophysical and geochemical prospecting. In regions where laterites and lateritic soils develop *in situ* on weathered bedrock, dispersion patterns of elements in the superficial cover can provide information about the rocks beneath. Geochemical surveying need not be confined to collection and analysis of stream sediment samples. Systematic soil surveys can also be made, though these have to be more carefully controlled, especially where thick laterites have developed and element dispersion is more extreme.

The engineering geology of superficial deposits and weathered bedrock can be very important in regions of low relief in low latitudes, such as West Africa. They are so thick and extensive that in many places they must support roads, dams and other structures. Because of the prolonged weathering under warm and humid climatic conditions, soil profiles are deep and generally rich in clay minerals, except where the bedrock consists of quartz-rich sandstones or metamorphic rocks. The high clay content of many West African laterites and soils makes them a particularly difficult problem in road and railway engineering, as well as in building construction. During rainy periods they may absorb large quantities of water, which causes them to expand and to become well lubricated. The expansion alone can cause fracturing in the foundations of buildings, but the greatest damage is done to roads and railways where stresses caused by the traffic cause roadbeds to shift and buckle. Engineers can carry out soil surveys when designing roads and railways and can site the routes over soils which are the best drained and have the lowest clay content.

Excavating to bedrock is an obvious alternative, but is often not feasible on account of the great thickness of the superficial cover. Moreover, it may not always be the best solution, especially where road cuts or other steep slopes are involved. In temperate climates, steep slopes cut into, for example, massive granites, are commonly considered to be safe engineering. In humid tropical climates, such rocks can weather quickly on exposure, the feldspars, micas and amphiboles turning quite rapidly into clays. Rocks containing high proportions of these easily weathered minerals can become unstable surprisingly quickly, and deep cuts with steep slopes may in the long term require costly measures to deal with weathering and slope failure.

Problems of comparable magnitude but of a different kind are met in civil engineering works along the coast, where there are many important harbours and industrial complexes as well as fisheries, not to mention recreation and tourist centres. The southern coastline of West Africa has been described as an almost continuous complex of sandy beaches, sand bars and lagoons. The sand is supplied partly by coastal erosion and partly by rivers in the wet season. The prevailing swell is from the south-west, so longshore drift is from west to east and sand bars grow eastwards. Engineering projects can disrupt the natural balance by changing either the pattern of longshore drift or the discharge of rivers to the sea (or both), and thus either create lagoons where none previously existed or breach and effectively destroy existing ones.

18.4.1 Water resources

Supplies of surface and near-surface water fluctuate with wet and dry seasons, particularly in the drier regions. Dams have been and are being built in many places to store water for dry season use. The reservoirs also provide water for irrigation projects and can support fishing industries. The three largest dams in West Africa were constructed primarily for hydroelectric power generation. The enormous man-made lakes formed by them, in Ivory Coast, Ghana and Nigeria (Bandana Blanc reservoir, Lake Volta and Kainji reservoir), have changed the face of the surrounding countryside. One problem with reservoirs in areas of seasonal rainfall is that evaporation during long dry seasons can greatly reduce the volume of water in them, and shortages can result, if rainfall is inadequate for a year or two.

Limited supplies of water are available in the superficial deposits and alluvium of river valleys and

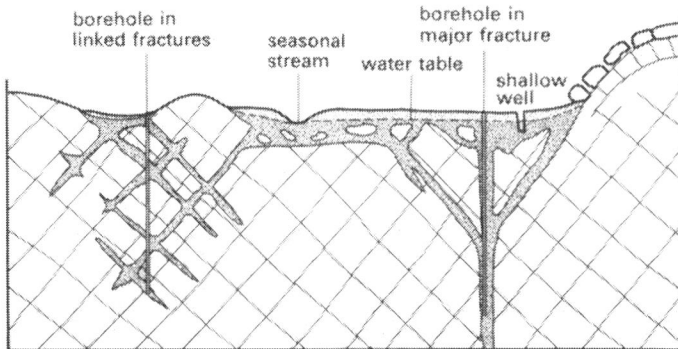

Figure 18.2 Diagrammatic cross section with exaggerated vertical scale to illustrate aspects of ground water in areas of crystalline rocks.

can be tapped by shallow wells dug through the water table. The water table can fall many metres during dry seasons, particularly on interfluves between streams, but even in the driest areas there is usually some water just below the surface in the sandy beds of larger seasonal rivers. The construction of dams helps to reduce the rate of lowering of the water table during dry seasons.

Perched water bodies can occur above the regional water table, for example in dune ridges, or in young volcanic rocks where scoriaceous or pyroclastic layers alternate with lava flows. Such aquifers are usually small but may contain enough water for small-scale local needs.

In basement areas, larger and more permanent supplies of ground water are stored in faults and fracture systems. These have been progressively widened with time by chemical weathering, as surface water percolates into them, and they are filled with permeable weathered rock debris. They can hold large quantities of water, as the larger fractures can be many kilometres long and extend to several hundred metres depth. Linked networks of smaller fractures can also provide substantial quantities of

ground water. These relationships are summarised in Figure 18.2.

Warm springs with water temperatures of about 35°C are known from at least two places (both in Nigeria), and they provide evidence that the water must have circulated to depths of at least several hundred metres, even allowing for a fairly steep thermal gradient.

Many of the sandstones and limestones in the sedimentary basins are good aquifers and as the sediments are sometimes permeable to considerable depths there may be large amounts of water stored beneath the water table, though the water table itself may be many tens of metres below the ground surface. If topographic depressions intersect the water table in such areas, perennial springs and pools or shallow lakes may occur, and water table wells can yield large quantities of water. Where layers of impermeable shale or clay alternate with permeable beds in gently dipping sediment sequences, there are confined aquifers containing artesian ground water. These relationships are summarised in Figure 18.3. The Iullmedden and Chad Basins are good examples of areas where both

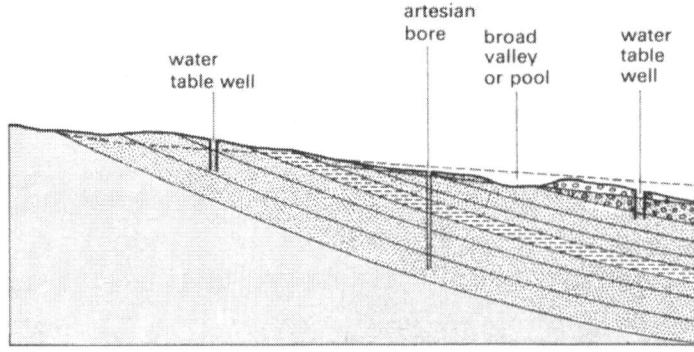

Figure 18.3 Diagrammatic cross section with exaggerated vertical scale to illustrate aspects of ground water in confined (artesian) and unconfined aquifers in sedimentary sequences.

unconfined and artesian aquifers occur, and their hydrogeology has been the subject of considerable research and development.

Most ground waters are of low salinity, though brackish waters with 500 mg l^{-1} dissolved salts or more are not uncommon. In parts of the Benue Trough there are brine springs, from which *salt* is locally extracted by evaporation. The origin of the brines is not known, but it has been suggested that they might be descendants of the hydrothermal solutions that gave rise to the lead–zinc mineralisation of the Benue Trough (Sec. 12.2.2). Salt springs are also reported from parts of the Senegal Basin.

Unlike other geological resources, supplies of water are mostly renewed each year by seasonal rainfall. Deeper ground waters may not be so easily replenished, however. Water in some of the artesian aquifers may be as much as 20 000 years old and, as rates of recharge are very low, the rate of extraction must be managed carefully to prevent excessive depletion. In northeastern Nigeria, for example, wells tapping water from several hundred metres depth were fountaining at the surface in 1968, but had ceased to flow ten years later.

Conservation of water supplies is one of the more important requirements in maintaining and improving the living conditions of the population in arid regions. Water is the only geological resource that is essential to life.

Bibliography

Akpati, B. N. 1978. Geologic structure and evolution of the Keta Basin, Ghana, West Africa. *Geol. Soc. Am. Bull.* **89**, 124–32.

Anderson, M. M. and W. D. Bruckner 1957. Note on raised shorelines of the Gold Coast. *Proc. 3rd Pan African Congr. on Prehistory*, No. 10, pp. 86–92.

Barber, W. 1965. *Pressure water in the Chad Formation of Bornu and Dikwa emirates, north-eastern Nigeria.* Geol. Surv. Nigeria, Bull. No. 35.

Beaudet, H., R. Coque, P. Michel and P. Rognon 1977. Altérations tropicales et accumulations ferrugineuses entre la vallée du Niger et les massifs centraux sahariens (Aïr et Hoggar). *Z. Geomorph.* **21**, 297–322.

Bondesen, E. and A. F. J. Smit 1972. Holocene tectonic activity in West Africa dated by archaeologic methods (discussion of paper by Thompson). *Geol. Soc. Am. Bull.* **83**, 1195–6.

Bonvallot, J. and B. Boulange 1972. Note sur le relief et son évolutions dans la région de Bongouanou, Côte d'Ivoire. In *African geology*, T. F. J. Dessauvagie and

A. J. Whiteman (eds). Ibadan: Ibadan University Press.

Brown, C. and R. W. Girdler 1980. Interpretation of African gravity and its implication for the breakup of the continents. *J. Geophys. Res.* **85**, 6443–55.

Burke, K. C. 1976. The Chad Basin: an active intra-continental basin. *Tectonophysics* **36**, 197–206.

Burke, K. C. and A. B. Durotoye 1972. The Quaternary in Nigeria. In *African geology*, T. F. J. Dessauvagie and A. J. Whiteman (eds). Ibadan: Ibadan University Press.

Carmouze, J.-P. and G. Pedro 1977. Influence du climat sur le type de régulation saline du lac Tchad, et relation avec les modes de sédimentation lacustre. *Sci. Geol. Strasbourg* **39**, 33–49.

Dei, L. A. 1972. The central coastal plains of Ghana: a morphological and sedimentological study. *Z. Geomorph.* **16**, 415–31.

Dowling, J. W. F. 1966. The mode of occurrence of laterites in northern Nigeria and their appearance in aerial photographs. *Engng Geol.* **1**, 221–33.

Du Preez, J. W. 1956. Origin, classification and distribution of Nigerian laterites. *Proc. III Int. West African Conf.*, pp. 223–34. Lagos: Nigerian Museum.

Durotoye, A. B. 1976. Quaternary sediments in Nigeria. In *Geology of Nigeria*, C. A. Kogbe (ed.). Lagos: Elizabethan Press.

Faure, H. 1980. Late Cenozoic isostatic movements in Africa. In *Earth rheology, isostasy and eustasy*, N. A. Morner (ed.). London: Wiley.

Faure, H. and J.-Y. Gac 1981. Will the Sahelian drought end in 1985? *Nature* **291**, 475–8. Also reply to discussion, 1981, *ibid.* **293**, 414.

Ford, S. O. 1981. The economic mineral resources of the Benue Trough. *Earth Evol. Sci.* **1**, 154–63.

Furon, R. 1963. *Geology of Africa* (English edn). London: Oliver & Boyd.

Hilton, T. E. 1964. River captures in Ghana. *Bull. Ghana Geog. Assoc.* **9**, 13–24.

Hilton, T. E. 1966. The Accra plateau: landform of a coastal savanna of Ghana. *Z. Geomorph.* **10**, 369–86.

Matheis, G. 1980. Secondary geochemical dispersion and bedrock reflection in the tropical rain forest terrain. *Erzmetall.* **33**, 180–5.

Matheis, G. and M. Pearson 1982. Mineralogy and geochemical dispersion in lateritic soil profiles of northern Nigeria. *Chem. Geol.* **35**, 129–45.

Mesida, E. A. 1978. Utilization of some lateritic clays for burnt bricks. *J. Min. Geol.* **15**, 108–14.

Michel, P. 1977. Recherches sur le Quatermaine en Afrique occidentale. *INOUA, Suppl. AFEQ Bull.*, 1977-1, **50**, 143–53.

Michel, P. 1978. Cuirasse bauxitiques et ferrugineuses d'Afrique occidentale. Aperçu chronologique. *Trav. Doc. Geogr. Trop.* **33**, 11–32.

Nicholson, S. E. 1982. Pleistocene and Holocene climates in Africa (discussion of paper by Sarnthein *et al.*). *Nature* **296**, 779.

Nwajide, C. S. and M. Hoque 1979. Gullying processes in south-eastern Nigeria. *Nigerian Field* **44**, 64–74.

Olorunfemi, B. N. 1984. Mineralogical and physico-chemical properties of Niger Delta soils in relation with geochemical problems. *J. Afr. Earth Sci.* **2**, 259–66.

Otezie, G. E. 1976. The hydrogeology of the north-western Nigeria basin. In *Geology of Nigeria*, C. A. Kogbe (ed.). Lagos: Elizabethan Press.

Palutikof, J. P., J. M. Lough and G. Farmer 1981. Senegal River runoff. (Discussion of paper by Faure and Gac). *Nature* **293**, 414.

Perseil, E. A. and G. Grandin 1978. Evolution minéralogique du manganese dans trois gisements d'Afrique de l'Ouest: Mokta, Tambao, Nsuta. *Min. Dep.* **13**, 295–311.

Pugh, J. C. and L. C. King 1952. Outline of the geomorphology of Nigeria. *S. Afr. Geog. J.* **34**, 30–7.

Rackham, L. J. 1972. The planation surfaces of a part of the basement complex of northern Nigeria: an interim report. In *African geology*, T. F. J. Dessauvagie and A. J. Whiteman (eds). Ibadan: Ibadan University Press.

Rogers, A. S., A. M. A. Imevbore and O. S. Adegoke 1969. Physical and chemical properties of the Ikogosi warm spring, Western Nigeria. *J. Mining Geol.* **4**, 69–81.

Sarnthein, M., G. Tetzlarg, B. Koopmann, K. Wolter and U. Pllaumann 1981. Glacial and interglacial wind regimes over the eastern subtropical Atlantic and north-west Africa. *Nature* **293**, 193–6. Also reply to discussion, 1982, *ibid.* **296**, 779–80.

Smith, A. G. 1982. Late Cenozoic uplift of stable continents in a reference frame fixed to South America. *Nature* **296**, 400–4.

Thomas, M. F. 1974. *Tropical geomorphology*. London : Macmillan.

Thompson, T. F. 1970. Holocene tectonic activity in West Africa dated by archaeologic methods. *Geol. Soc. Am. Bull.* **11**, 3759–64. Also reply to discussion, 1972, *ibid.* **83**, 1197–9.

Glossary

Abakaliki Basin Albian-Turonian basin at the lower end of the Benue Trough in which Asu River Group and Eze Aku and Awgu Shales accumulated

Abeokuta Formation Late Cretaceous non-marine and marine sediments in the eastern Dahomey Basin in southwestern Nigeria

aeolian drift Dust deposits carried over much of West Africa by northeasterly Harmattan winds during annual dry seasons, consisting mainly of quartz and clay minerals and redistributed by rain in wet seasons

African surface Erosion surface recognised by some geomorphologists throughout Africa and considered to be of late Cretaceous to mid-Tertiary in age, developed after the break-up of Gondwanaland. Identified as lying between about 500 and 800 m in parts of West Africa.

Agbada Formation Alternating sands and shales of the offshore (continental shelf) environment of the Niger Delta Basin. Strongly diachronous, becoming younger southwards

age province A region of the Earth's crust in which the ages of most of the rocks can be shown to lie within the same range of geological time

agmatite Normally applied to the structure formed when an amphibolite lens or layer has been disrupted into irregular blocks that are surrounded and intensively veined by pegmatitic material

Akata Formation Mainly muds and shales representing the delta front and continental slope environment of the Niger Delta Basin. Strongly diachronous, becoming younger southwards

Akroso Conglomerate Name given in Ghana to what may be the basal conglomerate of the Middle Voltaian (Pendjari Group). A fluvioglacial origin has been proposed for it, but it has not positively been identified as a tillite

Akwapimian An alternative name for the Togo Formation of the Togo belt

alkali basalt A variety of basaltic rock characterised by relative enrichment in alkalis, and typical of anorogenic rift and dome environments on continents, and of some oceanic islands. Commonly associated with phonolites and trachytes

Anambra Basin Upper Senonian-Maastrichtian and Palaeocene basin at the lower end of the Benue Trough in which Nkporo Shales and younger sediments accumulated, and which extended towards the southwest as the Niger Delta Basin

Anka belt A supracrustal belt in northwestern Nigeria characterised by minimal deformation and metamorphism of sediments and volcanics, and possibly analogous to the Tarkwaian rocks of the West African craton

anorogenic Any geological process that is not associated with orogenic (mountain-building) tectonic environments can be considered to be anorogenic

aplite Very fine-grained rock of granitic composition, commonly occurring as veins in association with granites and with pegmatites. Aplites are rarely repositories of economic minerals

Apollonian System The Mesozoic succession in the coastal Tano Basin of western Ghana, the eastern extension of the Ivory Coast Basin

Archaean Rocks with ages older than about 2500 Ma are assigned to the Archaean interval of geological time, the lower part of the Precambrian

Asu River Group Marine shales of Albian age in the southwestern part of the Benue Trough, deposited in the Abakaliki Basin

Atacorian or **Atacora Unit** An alternative name for the Togo Formation of the Togo belt

attapulgite A fibrous clay mineral with a chain silicate structure (as distinct from the sheet structure of most clay minerals). Its general formula is $(MgAl)_2Si_4O_{10}(OH).4H_2O$. Also called palygorskite

aulacogen An intracontinental rift that represents an abortive spreading axis, often the failed arm of a triple junction, where three such axes meet. The result is a rift-faulted depression extending inland from a continental margin along which lithospheric separation is taking place.

Awgu Shales Marine sediments of Upper Turonian to Lower Senonian (Coniacian) age in the southwestern (lower) segment of the Benue Trough

back-arc basin An elongate sedimentary basin floored either by oceanic or continental crust (ensimatic or ensialic respectively), developed above a subduction zone behind an active island arc or a growing continental margin mountain belt. Back-arc basins result from tensional stresses in the crust above active subduction zones, and if they lie between two island arcs, they are then also inter-arc basins

Badagba Quartzites Narrow elongate ridges of mainly quartzitic rocks in Benin, probably supracrustal strips infolded into the basement. They may be outliers of the Togo Formation and equivalent in age to the Younger Metasediments of Nigeria. The type locality is in southern Benin

Bakel Group Sediments and volcanics in eastern Senegal, of Infracambrian to Lower Palaeozoic age but possibly including older rocks, which are broadly equivalent to the Mali Group and are part of the main Mauritanide belt

Bama Ridge Prominent sand ridge running NW–SE through Maiduguri in northeastern Nigeria. It is a beach ridge, marking part of the shoreline of the ancient precursor of the present Lake Chad (Megachad)

banded iron formation (itabirite) Ferruginous quartzites consisting of alternating quartz-rich and haematite- or magnetite-rich layers or bands, rich enough in iron to be considered as potential sources of iron ore

Bandiagara Sandstones Cambrian(?) sediments of fluvio-deltaic origin unconformably overlying (with the Koutiala Sandstones) sediments of Supergroup 1 in the eastern part of the southern Taoudeni Basin

Baoulé–Mossi domain The Lower Proterozoic age province of the West African craton. Named after the Baoulé River in southern Mali and the Mossi region of central Upper Volta

Baoulé type granite See Cape Coast type granite

basement Used in two ways: *Either* in general, to denote the regional metamorphic and granitic rocks (mostly Precambrian) that underlie the generally flat-lying sediments – mostly Phanerozoic. *Or* applied in a specifically Precambrian context, to denote the high-grade regionally metamorphosed gneisses and migmatites that surround the elongate belts of lower-grade metamorphic rocks of supracrustal belts

basement complex A commonly used synonym for basement

Bassaris belt A name given to the northern continuation of the Rokelide belt, the segment between the northern margin of the Bové Basin and the junction with the main Mauritanide belt in southeastern Senegal and northern Guinea

bauchite A distinctive coarse-grained monzonite with fayalite and hypersthene that occurs as large intrusions associated with Older Granites in north-central and northeastern Nigeria. Bauchi is the type locality.

Benin Formation Mainly arenaceous sediments forming the upper layers of the Niger Delta Basin. Strongly diachronous, becoming younger southwards

Beyla Group Name applied to some of the mafic supracrustals (mainly amphibolites) in the Archaean terrane of Guinea

Bida (or Nupe) Sandstone Continental sands, grits and clays occupying most of the Bida Basin, of probable Campanian-Maastrichtian age

Bima (also Yola, Muri, Keana and Makurdi) Sandstones Massive cross-bedded feldspathic fluvio-deltaic sandstones of Albian-Cenomanian age occupying much of the middle and upper segments of the Benue Trough

Birimian Lithostratigraphic name given to the relatively low-grade metasedimentary and metavolcanic rocks west of the Togo belt and underlying much of Ghana, Ivory Coast and Upper Volta, as well as parts of Liberia, Guinea and southern Mali. The name derives from the original type locality of these rocks, the Birim River valley of Ghana. The rocks form the Lower Proterozoic part of the West African crater.

Bobo-Dioulasso embayment A southern extension of the Taoudeni Basin on to the West African craton, probably at the southern end of the Gourma aulacogen

Bondoukou type granite See Dixcove type granite

Buem Formation Sediments and volcanics occupying the western half of the Togo belt and the eastern margin of the West African craton and Volta Basin. The rocks dip generally eastward and are deformed mainly by thrusting. They are believed to be equivalent to the Middle Voltaian sediments of the Volta Basin. An alternative name for the formation is Thiélé Unit

Bullom Group Mainly Tertiary sands and clays with lignites forming the coastal strip of Sierra Leone

Burashika Group An assemblage of acidic igneous rocks in the upper Benue Trough of Nigeria, intrusive in the basement and of uncertain, but probably early Upper Cretaceous age.

Cameroun line The line of volcanic rocks and volcanoes that extends northeastwards from the volcanic islands of the Gulf of Guinea through Cameroun and into Chad

Cape Coast type granite Large concordant syntectonic batholithic granites of Eburnian age in Ghana. They are known as Baoulé type granites in Ivory Coast and Upper Volta

carbonatite A carbonate rock of magmatic origin, associated with alkaline to peralkaline igneous rocks and kimberlites. The main carbonate minerals are calcite and dolomite

cauldron subsidence The type of subsidence that is believed to occur beneath the calderas of large volcanoes, by a combination of basement fracture and evacuation of a magma chamber, leading to collapse of the superstructure into the resulting void or 'cauldron'

Chad Formation Quaternary fluviatile and lacustrine sands and clays (with diatomites) occupying most of the Chad Basin. Their maximum thickness is probably over 1000 m and they may extend down to the Pliocene

Coal Measures The Maastrichtian sandstones and shales with coal seams that overlie the Nkporo Shales at the southwestern end of the Benue Trough

coesite Dense polymorph of SiO_2 (2.93×10^3 kg m^{-3}), believed to be formed by meteorite impacts as it is found only in craters interpreted as being of meteorite impact origin

collophane A massive cryptocrystalline variety of apatite, the main constituent of most sedimentary phosphate rocks. The general formula is $Ca_3(PO_4)_2.2H_2O$.

cone sheets Thin sheets of igneous rock which have arcuate outcrop patterns and dip inwards towards a common focus a few kilometres below the surface

Continental Hamadien Non-marine sands and clays representing the Cenomanian to Senonian interval in the northwestern Iullmedden Basin

Continental Intercalaire Non-marine sediments covering the period between the late Carboniferous regression and the Upper Cretaceous transgression in northern Africa. As most of the sediments are of Jurassic to Lower Cretaceous age, the term is often taken to refer to this shorter interval

Continental Terminal   Non-marine sediments covering the period from the Lower Tertiary regression to the Quaternary, in northern Africa. It is a general term, as continental sediments of any age within this interval may be found in different places

craton   Area of continental crust stable since about 2000 Ma ago

cratonic nucleus   The oldest parts of cratons, regions geologically stabilised before the end of the Archaean, about 2500 Ma ago

crustal reactivation   (also called **reworking** or **remobilisation**)   When existing continental crust is caught up in orogenic deformation and regional metamorphism it is itself reheated and further deformed, with widespread resetting of radioactive 'clocks'

Dahomey Basin   The Mesozoic-Tertiary coastal basin straddling the Nigeria–Benin border, and extending south-west into Ghana as the Keta Basin

Dahomeyan   Lithostratigraphic name given to the metamorphic rocks east of the Togo belt and underlying much of Togo, Benin and Nigeria, as well as parts of Cameroun and Chad. The name derives from Dahomey, the earlier name for the Republic of Benin. Its use should be confined to the high-grade gneiss–migmatite basement that surrounds the lower-grade metasedimentary belts. Hower, it is often used to denote the whole metamorphic and igneous Pan African terrane

Damergou gap   The region between the Zinder inlier in southern Niger and the Aïr of the southeastern Hoggar, where the Iullmedden Basin is linked to the Chad Basin further east

Dange Formation   Marine shales and limestones belonging to the Sokoto Group of the Sokoto Basin

Dapango–Bombouaka Group   A new name for the Lower Voltaian sediments of the Volta Basin. Equivalent in age to the Togo Formation of the Togo belt

diachronous   When geological events such as marine transgressions, glaciations, orogenies and so on occur at different times in different places, they are said to be diachronous

diatreme   Explosive volcanic vent that pierces crustal rocks by high gas pressures, rapidly 'drilling' its way towards the surface

Dixcove type granite   Small discordant late- to post-tectonic granite intrusions of Eburnian age in Ghana. It is known as Bondoukou type granite in Ivory Coast and Upper Volta

domain   A region of the crust in which there is an overall similarity of lithology, structural style and trend, regional metamorphism and age. A larger age province can be subdivided into smaller domains, but an age province may itself constitute a domain

dropstones   Angular pebbles or boulders dropped from melting ice and depressing the lamination of soft muds into which they fall

Dukamaje Formation   Shallow-water marine shales belonging to the Rima Group of the Sokoto Basin, rich in vertebrate remains and formerly known as *Mosasaurus* Shales

East Saharan craton   The eastern Hoggar region of Algeria is believed to be composed of pre-Pan African rocks and appears to have been largely unaffected by the main Pan African deformation

Eburnian   Thermotectonic event, *c*. 2000 Ma ago, responsible for stabilising most of the West African craton

eclogite   Metamorphic rock composed mainly of garnet and clinopyroxene (jadeite), chemically equivalent to basalt and formed under very high-pressure conditions

Edina Sandstone   Mainly terrestrial sandstones of indeterminate Tertiary age outcropping on the coast of Liberia

Effon Psammite Formation   Name given to quartzites and quartzitic schists forming a distinctive supracrustal belt some 180 km long just to the east of Ilesha, in Nigeria

ensialic   Normally applied to depositional basins developed on continental crust. Such crust has been called sialic, from the term 'sial', indicating abundance of the chemical elements SIlicon and ALuminium

ensimatic   Normally applied to depositional basins developed on oceanic crust. Such crust has been called simatic, from the term 'sima', indicating abundance of the chemical elements SIlicon and MAgnesium

epeirogenic   Warping, differential uplift and faulting of tectonically stable crust are referred to as epeirogenic movements

Eze Aku Shales   Marine sediments of Turonian age in the southwestern (lower) segment of the Benue Trough

Falémé Group   A name given to lateral equivalents of the Mali Group that lie upon the craton to the east of the northern segment of the Rokelide belt (sometimes called the Bassaris belt), between the Bové Basin and the main Mauritanide belt. It represents Supergroup 2 of the Taoudeni Basin in the extreme west, adjacent to the Mauritanide belt

Farmington River Formation   Mainly terrestrial greywackes and conglomerates of indeterminate Upper Cretaceous age outcropping on the coast of Liberia

fenitisation   A form of metasomatic alteration of wall rocks by mainly alkali elements, accompanied by some desilication, in the vicinity of kimberlite and carbonatite intrusions

Fika Shales   Lateral equivalent of the upper part of the Pindiga Formation north of the Zambuk Ridge in the Benue Trough, overlying the Gongila Formation

fluidised system (**fluidisation**)   In geological contexts, the term describes hot gases streaming through fissures in rocks, fragmenting and pulverising them and transporting the fragments in suspension as a kind of fluid

Fluvio-Volcanic Series Sands, gravels and clays with kaolinised basalt flows, forming scattered outcrops on the Jos Plateau of Nigeria. Their age is uncertain, but is generally thought to be Lower Tertiary

Freetown complex A layered gabbro–anorthosite intrusive complex, forming the Freetown peninsula of Sierra Leone. It is of Triassic to lowermost Jurassic age

fuchsite Bright green mica related to muscovite, but with chromium (Cr^{3+}) replacing some aluminium (Al^{3+}) in the structure. General formula $K(Al,Cr)_2(AlSi_3O_{10})(OH)_2$

Gamba Formation Marine shales (laminated and in part phosphatic) belonging to the Sokoto Group of the Sokoto Basin

Gao Trough A narrow fault-bounded depression, filled with sediments and lying within the southern part of the Sudanese Strait. It underlies the northernmost part of the middle Niger valley.

Gombe Sandstone Non-marine sandstones overlying the Pindiga Formation and related formations in the upper (northeastern) segment of the Benue Trough, and probably of Maastrichtian age

gondite Manganese-rich rock composed of quartz and the Mn-garnet spessartite (spessartine), sometimes in alternating bands and thus approximating to a manganiferous analogue of banded iron formation. Supergene alteration of these rocks can lead to important concentrations of manganese oxide ores

Gondwana Surface Erosion surface recognised by some geomorphologists throughout Africa and considered to be of late Jurassic age, pre-dating the break-up of Gondwanaland, and generally higher than 1000 m above sea level

Gongila Formation Lateral equivalent of the lower part of the Pindiga Formation north of the Zambuk Ridge in the Benue Trough, probably mainly Turonian in age

granite–greenstone association Many Archaean terranes in all parts of the world consist typically of a granitic and high-grade gneiss–migmatite basement in which are elongate supracrustal belts consisting mainly of basic metavolcanic rocks of greenschist grade. These greenstone belts appear typically to occur as steep synformal infolds into the surrounding granitic rocks

greenstone belts Supracrustal belts of usually Archaean age that are made up largely of low-grade metavolcanic rocks of basaltic composition

greisen A late stage igneous or hydrothermal vein rock dominated by quartz and mica, often mineralised, especially with cassiterite and/or wolfram, and containing a variety of other accessory minerals, such as topaz, fluorite and sulphides

growth faults Penecontemporaneous normal (tensional) faults developed in thick continental shelf and deltaic sediment sequences, largely as a result of gravitational instability and slumping. Growth faults and associated anticlinal structures are important hydrocarbon trap structures in sediment sequences of this kind

Guinea Gneiss Name given to basement gneisses and migmatites in northwestern Guinea

Gundumi Formation Cretaceous non-marine sands and clays in the Sokoto Basin, probably mainly equivalent to the Continental Hamadien. Laterally equivalent to the Illo Formation

Hercynian Name given to the late to post-Carboniferous orogeny that affected much of continental Europe, but had only limited influence in Africa, where it formed part of the Mauritanide belt in the west. Another name for Hercynian is Variscan

high-level laterite Flat-topped erosion residuals of the oldest preserved laterite layer in West Africa, perhaps formerly continuous over nearly the whole region, and possibly of indeterminate early to mid-Tertiary age

ignimbrite A compact pyroclastic rock formed by the solidification of volcanic fragments carried in suspension in hot volcanic gases erupted from a vent or caldera fissure. The overall density of the system is high, so these particle-laden gas clouds travel over the ground surface like very fast lava flows

Illo Formation Cretaceous non-marine sands and clays that are probably lateral equivalents of the Gundumi Formation in the Sokoto Basin

imbrication Steeply inclined overlapping arrangement of thrust sheets, all subparallel and dipping in the same direction (also called imbricate structure)

impactite Vesicular glassy or finely crystalline rock produced by melting or partial melting as a result of the heat generated by meteorite impact

In Ouzzal block An elongate strip of probably Archaean crust caught up in the Pan African Pharusian belt of the Hoggar and separating its eastern and western branches

Infracambrian The uppermost part of the Proterozoic, immediately preceding the Cambrian and normally represented by unfossiliferous rocks. An alternative name is Vendian

inter-arc basin A normally elongate sedimentary basin that lies between a pair of island arcs. Where there are several island arcs, as in the western Pacific, there may be several inter-arc basins, some of which will also be back-arc basins

Irhazer Clays Lower part of the Continental Intercalaire (*sensu stricto*), probably late Jurassic fluvio-lacustrine clays underlying the Tegama Sandstones in the northern Iullmedden Basin

Ivory Coast Basin The Mesozoic-Tertiary coastal basin of Ivory Coast, extending eastwards across the Ghana border as the Tano Basin

Kalambaina Formation Marine limestones, marls and

166

shales (locally cementstones) belonging to the Sokoto Group of the Sokoto Basin

Kandé–Boukombé Series A sequence of schists and phyllites forming part of the Togo Formation in the northern part of the Togo belt, near the Togo–Benin border

Kasila domain The southwest border zone of the Archaean cratonic nucleus in West Africa. It comprises the granulites of the Kasila Group and the lower-grade partly overfolded Marampa Group supracrustals, and has a generally NW–SE structural grain

Kasila Group High-grade greenstone-type supracrustal belt in the Archaean(?) of western Sierra Leone, characterised by granulite facies metamorphism and a thrust-faulted contact with the cratonic basement. Together with the Sula and Marampa Groups it is part of the Kambui Supergroup

Katangan Name given to the inferred Upper Proterozoic age span of supracrustal sediments and volcanics that were metamorphosed and deformed to form the Pan African supracrustal belts of Nigeria

Keana Sandstone *See* Bima Sandstone

Kenema–Man domain This domain comprises most of the West African Archaean cratonic nucleus, with greenstone and banded iron formation rocks among the supracrustal belts, and an overall northerly to north-easterly structural grain

Kerri Kerri Formation Fluvio-lacustrine sediments of Palaeocene age on the western flank of the upper (north-eastern) segment of the Benue Trough

Keta Basin The extreme southwestern extremity of the Dahomey Basin, in Ghana, where the Mesozoic succession is known from boreholes to be underlain by Lower Palaeozoic sediments

Kibaran Thermotectonic event, *c.* 1100 Ma ago, which may have affected supracrustal metasediments and metavolcanics east of the West African craton. The evidence for this is patchy and inconclusive

kimberlite An alkaline peridotite with phenocrysts of often serpentinised olivine, chloritised phlogopite, and garnet (pyrope), in a fine-grained groundmass of magnesian ilmenite, serpentine, chlorite, magnetite and perovskite. The rocks show evidence of an origin deep in the Earth's mantle and are typically emplaced as pipe- or funnel-shaped bodies or as dykes, often with signs of explosive activity accompanying emplacement (diatremes). Some kimberlites are diamond-bearing

Kirtachi Quartzite The equivalent of the Lower Voltaian sandstones and conglomerates at the northern extremity of the Volta Basin in southwestern Niger

Koubia Group A sequence of Upper Proterozoic eugeosynclinal sediments and volcanics underlying the Mali Group, along with the Kerméssé Group, both probably lateral equivalents of the Madina-Kouta Group on the West African craton

Koulountou belt The southwestern branch of the Mauritanides, formed of rocks generally similar to those of the Mali Group in the Bassaris belt

Koutous Formation Late Cretaceous non-marine sands and clays in southern Niger, probably of Senonian-Maastrichtian age and equivalent in age to the Rima Group of the Sokoto Basin

Lafia Sandstones Mainly continental sediments of probable Campanian-Maastrichtian age occupying the northern side of the Benue Trough east of the Niger–Benue confluence. Probably equivalent to the Nkporo Shales and Coal Measures and to the Lokoja and Bida Sandstones

Leonian Thermotectonic event, *c.* 3000–2800 Ma ago, for which there is increasing evidence in rocks of the West African cratonic nucleus

Liberian Thermotectonic event, *c.* 2750 Ma ago, responsible for stabilising the Archaean nucleus of the West African craton

Lokoja Sandstone Continental sands, grits and clays near the Niger–Benue confluence, lateral equivalents of the Lafia Sandstone to the east and the Bida (Nupe) Sandstone to the north

'low' and 'high' Africa The northern and western parts of the African continent lie mostly below 1000 m altitude ('low' Africa), in contrast to the southern and eastern parts which lie mostly above this level ('high' Africa)

Macenta Gneiss Name given to basement gneisses and migmatites in Guinea west of the Simandou belt

Madina–Konta Group Upper Proterozoic platform sediments, mainly sandstones and shales, possibly lateral equivalents of the Koubia and Terméssé Groups forming the lower part of the Bassaris belt

Mahana Gneiss Name given to basement gneisses and migmatites in Guinea east of the Simandou belt

Makurdi Sandstone *See* Bima Sandstone

Mali Group Sediments and volcanics of Infracambrian to Lower Palaeozoic age in southeastern Senegal and northern Guinea, correlated with the Rokel River Group and also deformed and metamorphosed in the Pan African

Man charnockite complex Massif in western Ivory Coast formed of charnockitic granites, noritic gabbro–anorthosite, and amphibolites and pyroxenites, as well as high-grade metasediments (magnetite quartzites) and gneisses

Marampa Group Low-grade belt of greenstone-type belt supracrustals in the Archaean(?) of western Sierra Leone, characterised by recumbent folding. Together with the Sula and Kasila Groups it is part of the Kambui Supergroup

Megachad The giant shallow lake that once occupied a much greater area of the Chad Basin than the present lake, about 10000 years ago

metallogenic province A region of the crust that is especially well endowed with metalliferous mineral deposits. The area of a metallogenic province may be anything from a few thousand to a few hundred thousand square kilometres

metasediments Simply an abbreviation for metamorphosed sediments. Usually refers to relatively low-grade rocks in which original sedimentary features can still be recognised

metavolcanics Simply an abbreviation for metamorphosed volcanics. Usually refers to relatively low-grade rocks in which the original volcanic (or at least igneous) nature of the rocks can be established with some confidence. Metamorphosed basic rocks are also called metabasic rocks or metabasites

Migmatite and Gneiss Formation The main area of Archaean basement in western Ivory Coast, probably younger than the higher-grade Mt Douan Formation and the Man charnockite complex

mineralogenic province A region of the crust that is especially well endowed with non-metalliferous mineral deposits (for example, phosphate, coal or petroleum), though metalliferous deposits are not excluded. The area of a mineralogenic province may be anything from a few thousand to a few hundred thousand square kilometres

mixtites A general term applied to unsorted bouldery deposits that may be of glacial origin but could also be the products of debris flows or turbidity currents

mobile belt Regions of continental crust that are younger than the cratons. They experienced orogenic deformation and regional metamorphism during the last 1500 Ma or so

Monrovia Diabase A dolerite sill of probable Jurassic age intruding the Paynesville Sandstone of Liberia. Diabase is an alternative (American) term for dolerite

Mosasaurus Shales An early name for the Dukamaje Formation of the Sokoto Basin

Mt Douan Formation Granulites and charnockitic gneisses occupying an antiformal structure surrounded by migmatites in the Kenema–Man domain of western Ivory Coast

Mt Gao Formation The Archaean supracrustal assemblage of quartzites, ironstones and amphibolites, amphibole–pyroxenites and garnet–pyroxenites in western Ivory Coast

Muri Sandstone *See* Bima Sandstone

Nauli Limestone Maastrichtian limestone forming a ridge in the onshore succession of the Tano Basin

Ndias dome Broad anticlinal structure in Cretaceous sediments forming much of the Cape Verde peninsula east of Dakar in Senegal

Ngaoundere Fault Zone Major fault system in Cameroun, probably originating in the Pan African and rejuvenated in both the Cretaceous and Tertiary. It is marked by major zones of mylonite

Nimba Group A name for the Archaean supracrustal greenstone–banded iron formation belts of Liberia, mostly in the western half of the country. The name derives from one of the largest, the Nimba belt on the Liberia–Ivory Coast–Guinea border, containing very rich banded iron formation. (Sometimes called the Nimba Series)

Nkporo Shales Upper Senonian (Campanian) marine sediments, including limestones and massive sand stones, in the southwestern (lower) segment of the Benue Trough

nsutite A hydrated manganese oxide mineral, formerly called gamma-MnO_2, with the formula: $Mn^{4+}_{1-x}Mn^{2+}_x O_{2-2x}(OH)_{2x}$. The type locality is Nsuta, southern Ghana, after which the mineral is named. Large deposits of this mineral were once mined to produce battery-grade MnO_2, but they are now exhausted

Nupe Sandstone *See* Bida Sandstone

Obosum Formation Sediments of the Upper Voltaian, derived from erosion of the Togo belt

Odukpani Formation Arenaceous marine sediments of Cenomanian age outcropping over a restricted area at the southwestern end of the Benue Trough

Older Granites The syntectonic to late-tectonic intrusions of Pan African age in Nigeria are given this name to distinguish them from the Jurassic Younger Granites of the Jos Plateau

Older Metasediments Supracrustal relics of possible Eburnian or greater age in the Dahomeyan basement of Nigeria. Mostly quartzites, marbles and amphibolites; larger bodies are not always easy to distinguish from Younger Metasediments

orthogneiss Name given to high-grade metamorphic rocks that can reasonably be inferred to be of igneous parentage, such as granitic gneisses, and amphibolites

Oti Formation A name given to the Middle Voltaian in Ghana

Pan African In relation to African geology, this term refers to rocks that experienced orogenic deformation and regional metamorphism in the period *c.* 650–450 Ma ago

paragneiss Name given to high-grade metamorphic rocks of sedimentary origin, such as many biotite- and muscovite-rich gneisses, quartzites, marbles, calc-silicate rocks and so on

Paynesville Sandstone Palaeozoic sandstones of terrestrial origin outcropping on the Liberian coast between Monrovia and Buchanan, and reaching an estimated total thickness of 1000 m

pediment Broad flat or gently sloping erosion surface developed under semi-arid to arid conditions and generally backed by hills or mountains

pegmatite Very coarse-grained rock of mainly quartz–feldspar composition, commonly occurring as veins or pods in association with granitic rocks. Some pegmatites are rich sources of economic minerals, but the majority are barren

pelitic Term applied to metamorphic rocks of originally clay-rich (argillaceous) sedimentary origin. Meta-argillites is a term that is also sometimes used for less strongly metamorphosed pelitic rocks, or pelites

Pendjari Group A new name for the Middle Voltaian sediments of the Volta Basin. Also called the Oti Formation in Ghana. Equivalent in age to the Buem Formation of the Togo belt

peneplain Extensive plain of low relief developed by long-continued erosion to near the base level of major rivers, generally developed in humid climatic conditions

peralkaline Geochemical term applied to rock analyses in which there is an excess of $K_2O + Na_2O$ (total alkalis) over and above that required to form feldspars with the available silica and alumina (SiO_2 and Al_2O_3), so that either feldspathoids or alkali-rich ferromagnesian minerals are formed

peraluminous Geochemical term applied to rock analyses in which there is an excess of Al_2O_3 (alumina) over and above that required to form feldspars with the available SiO_2 and alkalis, so that corundum appears as a phase in the norm

petrographic province An assemblage of (usually) igneous rocks with an overall similarity in terms of petrography and geochemistry, emplaced in a particular geotectonic setting within a reasonably well defined area and age range

Phanerozoic A general name for the Cambrian to Recent interval of geological time, comprising rocks that can be dated by fossil remains

Pharusian belt The Pan African belt of the western Tuareg shield, made up of Upper Proterozoic rocks, metamorphosed, folded and thrust on to the West African craton

Pindiga Formation Marine sediments (shales with limestones) of certain Turonian age in the upper (north-eastern) segment of the Benue Trough, which may range up to Maastrichtian in age, or may be merely Turonian-Coniacian

polycyclic In general, the term means 'subjected to many cycles'. Used in connection with basement terranes in particular where there is structural and geochronological evidence that repeated reactivations of continental crust have occurred

Porphyritic Older Granite Name given to a distinctive adamellitic variety of Older Granite in Nigeria, characterised by the occurrence of large microcline megacrysts

Proterozoic Rocks with ages in the range c. 2500–600 Ma are assigned to the Proterozoic interval of geological time, the upper part of the Precambrian

psammitic Term applied to metamorphic rocks of originally sandy quartzose, arkosic or greywacke (arenaceous) sedimentary origin. The rocks are also called psammites. Meta-arenite is a term that is also sometimes

used for less strongly metamorphosed psammitic rocks, or psammites

relict ages Heating that accompanies regional metamorphism during orogenic events can reset the radioactive 'clock' systems that are used to date rocks. Most easily reset is the K/Ar system. The Rb/Sr and U/Pb systems can be variably reset, and the relatively recently developed Sm/Nd system is generally unaffected. Rocks can therefore contain relict radiometric evidence of earlier orogenic events to which they may have been subjected

ring complex Igneous intrusions formed by a succession of roughly concentric and more or less cylindrical magma bodies emplaced in anorogenic environments, mostly in continental crust. The rocks may range in composition from gabbro to granite, syenite and nepheline syenite

ring dyke A thick cylindrical intrusion of igneous rock, typically with steep sides

roche moutonnée Rock outcrops that have been underneath a glacier and have been both smoothed (upstream) and broken (downstream) by the moving ice. So-called because from a distance they resemble flocks of sheep

Rokel River Group The sediments and minor volcanics that occur within the Rokelide belt, ranging in age from late Proterozoic to lowermost Phanerozoic

Rokelide belt A narrow supracrustal belt on the western margin of the West African craton, mostly in Sierra Leone but extending north-west into Guinea. Characterised by relatively weak deformation and metamorphism in the Pan African, with minor reactivation of Archaean cratonic rocks. It is a southern branch of the Mauritanide belt

Sahelian belt Arid to semi-arid zone extending from 15°N latitude to the Sahara desert. Also known simply as the Sahel

Saionia Scarp Group Ordovician sediments, in part of fluvioglacial origin, representing the top of Supergroup 2 and the lower part of Supergroup 3, and forming the lower part of the Bové Basin succession in Sierra Leone. Equivalent to the Youkounkoun Group further north.

Sasca domain An area in the southwestern part of the large Proterozoic Baoulé–Mossi domain in Ivory Coast where the rocks yield both Liberian and Eburnian ages. It could also be regarded as a southeastern part of the Archaean Kenema–Man domain

Sassandra mylonite zone A major N–S belt of shearing and mylonitisation that helps to define the eastern edge of the Archaean nucleus of the West African craton in western Ivory Coast. Named for the Sassandra River in SW Ivory Coast

Scolithus Generic name given to tubular and vermiform trace fossils, believed to have been formed by marine worms in shallow-water sediments of (mainly) Infracambrian age

screen A term applied to usually crescentic areas of basement that lie between adjacent ring dykes in ring complexes

série pourprée The reddish molasse deposits of the Pan African Pharusian belt in the Hoggar

série verte A name sometimes given to the 'greenstones' of the Pharusian belt of the Hoggar

shield Used in two ways: *Either* in general, to describe the whole of the tectonically stable Precambrian crust of the continent. *Or* applied in a more specific way to the cratons, i.e. to areas stabilised around 2000 Ma ago or more

silexite A name often applied by French writers to massive chemically precipitated cherts

Simandou Group A name for the Archaean supracrustal greenstone–banded iron formation belts of Guinea, mostly in the southern and western parts of the country. The name derives from the Simandou range. Alternative spellings are Siamandou and Sinandou. (Sometimes called Simandou Series)

Sokoto Basin The southeastern sector of the Iullmedden Basin

stromatolite Laminated structures of organic origin in limestones, representing ancient algal mats formed by growth of blue-green algae in very shallow water, usually under warm climatic conditions

Sudanese Strait (Détroit Soudanais) The narrow strip of Cretaceous to Tertiary marine sediments between the Adrar des Iforas and the West African craton, formerly a narrow seaway that gave access to the Iullmedden Basin

Suggarian belt The polycyclic terrane of the central Hoggar–Aïr segment of the Tuareg shield, which has similarities with the Dahomeyan/Katangan terranes of the eastern Pan African domain to the south

Sula Group A name for the Archaean supracrustal greenstone belts of Sierra Leone. There are nine such belts or areas of belt relics, and the name derives from one of the largest, the Sula Mountains belt. Together with the Marampa and Kasila Groups it is part of the Kambui Supergroup

Supergroup 1 The Upper Proterozoic sediments which form the lower part of the succession in the Taoudeni and Volta Basins, and are equivalent to the Lower Voltaian

Supergroup 2 The Infracambrian (Vendian) to Cambro-Ordovician sediments which form most of the upper part of the succession in the Taoudeni and Volta Basins. The base is marked by the triad, and the sediments are equivalent to the Middle Voltaian

Supergroup 3 The late Ordovician to Devonian and Carboniferous sediments forming the uppermost part of the succession in the Taoudeni and Volta Basins. The base is marked by a tillite and the sediments are equivalent to the Upper Voltaian

supracrustal belts Generally elongate zones of intensely folded but relatively low-grade metasediments and metavolcanics that are believed to have been deposited upon higher-grade basement and subsequently downfolded or downfaulted within it, during regional deformation and metamorphism which also reactivated the basement itself

Talach depression The northernmost part of the Iullmedden Basin, especially north-west of the Aïr, where both marine and non-marine Palaeozoic sediments were deposited. Also called the Tin Serririne syncline

Tano Basin The southeastern end of the Ivory Coast Basin, in Ghana

Tarkwaian A subordinate arenaceous and conglomeratic sequence of Lower Proterozoic age occurring among the Birimian supracrustals in Ghana, Ivory Coast and Upper Volta. There are some intrusive igneous rocks and these were folded and metamorphosed along with the sediments during the Eburnian event. Conglomeratic facies are gold-bearing in Ghana

Tegama Sandstone Group The youngest and most extensive member of the Continental Intercalaire in the Iullmedden Basin, mainly of Lower Cretaceous age

Tektites Small (*c.* 2–3 cm) glassy silica-rich bodies of various shapes believed to be solidified droplets of molten rock 'splashes' produced by high-velocity meteorite impacts

Terméssé Group A sequence of Upper Proterozoic eugeosynclinal sediments and volcanics underlying the Mali Group; along with the Koubia Group, both probably lateral equivalents of the Madina-Kouta Group on the West African craton

thermotectonic event A general term for the regional deformation and metamorphism that has affected large areas of continental crust, used when it cannot be unequivocally established that the processes were associated with those of a 'typical' mountain-building orogeny

Thiélé unit An alternative name for the Buem Formation of the Togo belt

tholeiitic basalt A variety of basaltic rock characterised by a relative deficiency in alkalis, and forming most of the topmost igneous layer of oceanic crust; also characteristic of basaltic magmatism in continental margin settings along the margins of opening oceans. Commonly associated with small volumes of rhyolitic rocks

tholoid A term applied to usually dome- or plug-shaped masses of igneous rock that have been extruded at the surface as viscous magma, the viscosity being too high for the magma to form a 'normal' lava flow. Tholoids are normally formed by intermediate to acid rocks such as trachytes and rhyolites

Tichit Group Ordovician sediments of glacial and fluvioglacial origin that form the base of Supergroup 3 in the southern Taoudeni Basin

tilloids A general term applied to unsorted bouldery deposits which resemble tillites and for which a glacial origin is suspected but not unequivocally indicated by the evidence

Tin Serririne syncline The northernmost part of the Iullmedden Basin, especially northwest of the Aïr, where both marine and non-marine Palaeozoic sediments were deposited. Also called the Talach depression

toe thrusts Penecontemporaneous reverse (compressional) faults developed at the seaward end of thick continental shelf and deltaic sediment sequences, largely as a result of gravitational instability and slumping

Togo belt Strongly compressed and thrust-faulted supracrustal sediments and volcanics of probable Katangan age, occupying a broad NNE–SSW trending strip of country between the eastern Pan African domain and the West African craton

Togo Formation Metamorphosed sediments and volcanics occupying the eastern half of the Togo belt and adjoining the Dahomeyan terrane of the eastern Pan African domain. The rocks dip generally eastwards and are deformed by folding and thrusting. They are believed to be equivalent to the Lower Voltaian sediments of the Volta Basin. Alternative names are Atacorian or Atacora Unit and Akwapimian

triad The association of tillite, stromatotillitic limestones with barite, and silexite, which characterises the Vendian glaciation interval in the Volta and Taoudeni Basins

Tuareg shield A name sometimes given to the Precambrian block of the Algerian Hoggar and its southern projections into Mali and Niger, the Adrar des Iforas and the Aïr. Also sometimes called the Hoggar shield

Type I and Type II belts Respectively greenstone-dominated and metasediment-dominated supracrustal belts in the Birimian of francophone West Africa

Vendian The last of the Precambrian intervals, the uppermost Infracambrian, *c.* 660–570 Ma ago

Wukari Formation Undifferentiated marine sediments occupying much of the middle sector of the Benue Trough, probably mainly shales, with interbedded limestones and sandstones

Yola Sandstone *See* Bima Sandstone

Yolde Formation Lowermost Turonian partly marine sediments representing a transition to fully marine conditions in the upper (northeastern) segment of the Benue Trough

Youkounkoun Group Ordovician sediments forming the lower part of the Bové Basin succession in Guinea and Senegal, equivalent to the Saionia Scarp Group in Sierra Leone

Younger Granites The anorogenic granite-dominated ring complexes that extend from Niger in the north to Cameroun in the south and range in age from Palaeozoic to Tertiary

Younger Metasediments Supracrustal belts of Katangan age in the Dahomeyan basement of Nigeria. Generally of lower metamorphic grade than the basement, but where grades are higher distinction from Older Metasediments cannot always be definitive

Zambuk Ridge A basement 'high' near the northern end of the Benue Trough, partly separating it from the Chad Basin

Zungeru mylonites A broad belt of recrystallised and deformed mylonitic rocks in central Nigeria, previously identified as metasediments and metavolcanics

Index of place names

The numbers in this index refer to text sections, except for occasional references to chapters. For features or regions that are also of geological or geographical significance, see the Subject index.

Subject index

The numbers in this index refer to text sections, rather than individual pages, *except* for occasional references to chapters and text tables (always prefixed by 'Table'). Entries shown in **bold** type correspond to terms explained in the Glossary, and the bold section number in the same index entry refers to the text section in which the term is introduced (or given extensive consideration).

175

176

177

178

179

183

Geology and mineral resources of West Africa

Reader survey

By completing and returning this questionnaire, you can help the authors and publishers to improve this book. We will make use of your replies when preparing the next edition.

Name and address (optional):

Which of the following main features of the book did you find:

1 very useful	2 fairly useful	3 not very useful		4 not at all useful	
subdivision into four main parts	1	2	3	4	
general study guide at the beginning	1	2	3	4	
summaries at the start of each chapter	1	2	3	4	
use of bold type to introduce terms	1	2	3	4	
glossary of terms	1	2	3	4	
sections on economic potential at the end of each chapter	1	2	3	4	
illustrations – maps and diagrams	1	2	3	4	

The following questions require brief explanatory comments. If we have not allowed enough space, please use the blank sheet at the end. If you have given your name and address, we can write and follow up your comments. Any advice or material we use will, of course, be acknowledged.

Which chapters or sections are too detailed for a book of this type?

Which chapters or sections are not detailed enough?

Which chapters or sections need up-dating?

Which chapters or sections are difficult to understand?

Which illustrations require modification or improvement?

Which illustrations should be removed or replaced?

What new material should be introduced?

What other improvements do you suggest?

Thank you very much for your help.

Please return your completed questionnaire to:

Dr J. B. Wright
Department of Earth Sciences
The Open University
Walton Hall
Milton Keynes MK7 6AA, UK

Printed in Great Britain
by Amazon